Debunking Seven Terrorism Myths Using Statistics

ASA-CRC Series on
STATISTICAL REASONING IN SCIENCE AND SOCIETY

SERIES EDITORS
Nicholas Fisher, University of Sydney, Australia
Nicholas Horton, Amherst College, MA, USA
Regina Nuzzo, Gallaudet University, Washington, DC, USA
David J Spiegelhalter, University of Cambridge, UK

PUBLISHED TITLES

Errors, Blunders, and Lies: How to Tell the Difference
David S. Salsburg

Visualizing Baseball
Jim Albert

Data Visualization: Charts, Maps and Interactive Graphics
Robert Grant

Improving Your NCAA® Bracket with Statistics
Tom Adams

Statistics and Health Care Fraud: How to Save Billions
Tahir Ekin

Measuring Crime: Behind the Statistics
Sharon Lohr

Measuring Society
Chaitra H. Nagaraja

Monitoring the Health of Populations by Tracking Disease Outbreaks
Steven E. Fricker and Ronald D. Fricker, Jr.

Debunking Seven Terrorism Myths Using Statistics
Andre Python

For more information about this series, please visit: https://www.crcpress.com/go/asacrc

Debunking Seven Terrorism Myths Using Statistics

Andre Python
University of Oxford
Oxford, United Kingdom
and
University of Zhejiang
Hangzhou, China

CRC Press
Taylor & Francis Group
Boca Raton London New York

CRC Press is an imprint of the
Taylor & Francis Group, an **informa** business
A CHAPMAN & HALL BOOK

First edition published, 2020
by CRC Press
6000 Broken Sound Parkway NW, Suite 300, Boca Raton, FL 33487-2742

and by CRC Press
2 Park Square, Milton Park, Abingdon, Oxon, OX14 4RN

© 2020 Andre Python
CRC Press is an imprint of Taylor & Francis Group, LLC

ISBN: 978-0-367-47228-3 (hbk)
ISBN: 978-0-367-47224-5 (pbk)
ISBN: 978-1-003-03423-0 (ebk)

Typeset in LMRoman
by Nova Techset Private Limited, Bengaluru & Chennai, India

**Visit the Taylor & Francis Web site at
http://www.taylorandfrancis.com**

**and the CRC Press Web site at
http://www.crcpress.com**

To my parents

Contents

Foreword

Debunking Seven Terrorism Myths Using Statistics constitutes a long-desired book for students, scholars, practitioners and a general public. Written in simple and delicate yet informative and well-sourced style, this book provides evidence-based findings on trends in terrorism, while reviewing and highlighting limitations imposed by theory, method and data. It provides to the reader a rigorous way of thinking and statistical tools necessary to develop features of a critical thinker. Hidden nuances, which one must keep in mind while considering questions about terrorism, are often overlooked. Ignorance to such details or lack of awareness could easily lead to overgeneralization, misinterpretation and false statements.

The book invites the reader to take part of a scientific journey, offering an accessible explanation of decisions made by scientific researchers, step by step revealing associated caveats. The book gradually prepares the reader to tackle more complex concepts and methods. It suggests a rigorous framework to study terrorism data within the context of methodological development, adopting scholarship from other disciplines. By doing so, this book demonstrates the benefits of evidence-based interdisciplinary research applied to the study of terrorism.

The book presents recent developments in the field of terrorism studies combining innovative concepts with advanced methods using data coming from the most relevant cases. It explains the use of statistical methods, conveniently indicating what questions can and cannot be answered by means of quantitative methods and what are possible limitations.

The book sets a stage of how to think statistically about the dependencies of terrorism in both space and time aiming at unpacking the complex dynamics of terrorism at relevant scales. The book translates statistical findings into a general language. It encourages the reader to be a critical consumer of reports about terrorism based on the analysis of data.

The informative visualization such as maps and graphs supplemented by the detailed explanation of how to interpret them significantly contributes to a smooth reading. Statistical graphs for a non-statistician might at times look unapproachable, and the effort that the author puts into their clear and simple explanation is very helpful.

The book demystifies at times trivialized statistical concepts such as prediction and educates a general readership of its true meaning in the context of statistics. Speaking 'the same language', being familiarized with foundational notions of statistics contributes to clarification of where and how statistical methods are at the core of modern terrorism studies.

This book is a must read for everyone who wishes to make an informed judgment on terrorism.

Dr. Elena Zhirukhina
Prague, 22 February 2020

Preface

Terrorism threatens our society in various parts of the globe. In the hope of explaining and preventing terrorist events, scholars have developed and applied statistical methods to analyze ever-growing databases on terrorism. Over the last decade, their work has generated unprecedented insight into terrorism. Yet their findings have not reached the public.

Debunking Seven Terrorism Myths Using Statistics is an attempt to communicate evidence-based research work on terrorism to a general audience. It illustrates key statistics that provide an overview of the extent and magnitude of terrorist events perpetrated by non-state actors across the world. This book describes the current state of knowledge on global patterns of terrorism and avoids jargon wherever possible.

While remarkable technical advances have been accomplished in the field of statistics and computer science along with the development of more comprehensive databases, the very nature of terrorism remains opaque due major theoretical and methodological limitations. Inspired by Plato's allegory of the cave, the book's cover image conveys the meaning of the discrepancies between light and shadow, reality and observations. I invite the readers to develop their own interpretations of the image before and after reading the book.

This book aims at providing a critical and nuanced view of the observed spatial and temporal patterns of terrorism obtained from the analysis of terrorism data. It makes the reader aware of the many obstacles that will ineluctably occur throughout the analysis of terrorism data. Furthermore, it provides guidance to mitigate them and interpret the findings within a coherent and rigorous methodological framework.

Andre Python
Oxford, 23 February 2020

Author

Andre Python holds a master's degree in Geography from the University of Fribourg, Switzerland. After completing a PhD in Statistics at the University of St Andrews, Andre pursued his academic career in the United Kingdom as a postdoctoral researcher in geospatial modeling at the University of Oxford before joining Zhejiang University (China) as a ZJU100 Young Professor of Statistics.

Andre has published his work on terrorism in peer-reviewed scientific journals, including *Journal of Conflict Resolution, Journal of the Royal Statistical Society - Series A : Statistics in Society* and *Political Geography*. Furthermore, he has disseminated his research work to a wide range of audiences through publications in popular science magazines such as *Significance* and various public talks given inside and outside academia.

His current research interests are in extending spatial models to address policy-relevant issues raised by the spread of phenomena threatening global security and health.

In line with the objectives and philosophy of the Royal Statistical Society, Andre believes that statistics and data have key roles in society. His work aims at providing evidence-based results to help decision makers use statistics effectively in the public interest.

Acknowledgment

This project would have been neither initiated nor completed without the kind and generous support from the Editor John Kimmel, the Series Editors, and the Reviewers. I am also very grateful to those who helped me get through this writing journey: Mike Thorn, Xavier Ruffieux, Jürgen Brandsch, Timothy Wilson, Richard English, Michela Cameletti, Nils Weidmann, Talip Alkhayer, Guido Sanguinetti, Timo Kivimaki, Manlio Fossati, and Elena Zhirukhina.

I would like to thank my academic mentors for their guidance and kind support: Martin Beniston, Gregor Kozlowski, Mathias Thoenig, Dominic Rohner, Janine Illian, Javier Argomaniz, Marta Blangiardo, Ewan Cameron, David Baker, and Katherine Battle. Thank you to my colleagues for their great support: Anita Nandi, Penny Hancock, Tim Lucas, Punam Amratia, Justin Millar, Rohan Arambepola, Andreas Bender, and Melchizedek Mashiku.

A special thank you to Baoli Liu for her advice and work on the graphics and her pleasant company throughout this adventure, which included book editing sessions between Moscow and Ulaanbaatar on the trans-Mongolian train! I am also infinitely grateful to the love and support I received from my siblings: Vladimir, Anne, and Martin and my cousin Mišo (and his famous "element of surprise theory"), his wife Mojca and their adorable cat Tobi.

I would like to conclude by expressing my gratitude to my mother for her unconditional love, courage and humility. She has conveyed her passion for the study of nature and stimulated my curiosity about science and its beauty. She remains my greatest source of inspiration.

CHAPTER 1

Introduction: The Role of Statistics in Debunking Terrorism Myths

> If it disagrees with experiment, it's wrong. In that simple statement is the key to science. It does not make any difference how beautiful your guess is, it does not matter how smart you are who made the guess, or what his name is.
>
> Richard Feynman

Terrorism threatens our society in various parts of the globe. In September 11, 2001, a series of coordinated terrorist attacks killed almost 3,000 people in the World Trade Center, New York and at the Pentagon. This notorious event, known as 9/11, became the most deadly terrorist attack perpetrated by non-state actors in world history. Since then, a vast amount of research work has been carried out to better understand the underlying mechanisms behind terrorism in the hope of preventing future potentially devastating acts of terror.

Recent advances in statistics and computer science along with an increasing availability of large databases have allowed scholars to develop and apply novel research methods to get unprecedented insight into terrorism. Yet these findings have not reached the public. This book is an attempt to communicate evidence-based research work to a general audience. By providing an illustrated description of a few key statistics that characterize terrorist events, it describes the current state of knowledge on key characteristics of terrorism free of scientific jargon that would only distract attention from the main thrust of recent findings. Simply put, this work is essentially about communicating knowledge on terrorism obtained from the analysis of terrorism data.

Throughout the book, seven terrorism myths will be debunked based on the analysis of terrorism data carried out in the R statistical software. R is an open-source programming language for data analysis, statistical computing, and graphics [116]. The use of a widely used software allows us to adopt a reproducible work flow so that: (a) the results can be easily reproduced, (b) readers can examine the decisions made to generate the results, and (c) one can reuse our methods in other contexts.

Chapter 2 highlights how different views on terrorism impact the observed spatial pattern of terrorist attacks. It informs the reader about the inherent ambiguity of the concept of terrorism and the absence of a consensual definition of terrorism. As an illustration, it highlights how diverging views of terrorism from the data providers influence the observed spatial patterns of terrorism data. Furthermore, it makes the reader aware of the effect of subsetting terrorism data into categories. In particular, it illustrates cases where the following classification of terrorist events is considered: (i) civilian versus non-civilian targets, (ii) state versus non-state terrorism, and (iii) political versus non-political terrorism.

Through an exploratory study of the *lethality*—the propensity to cause deaths—of terrorism, Chapter 3 demystifies the idea that terrorism is only about killing civilians. Based on empirical evidence gathered from large-scale datasets, it compares the proportion of attacks that killed people with those that do not provoke deadly casualties. Furthermore, it suggests alternative approaches to distinguish high casualty from low casualty attacks, and points out how each classification may lead to different interpretations. Moreover, this chapter suggests potential mechanisms that explain why some terrorist groups aim at inflicting large casualties while other groups avoid lethal casualties. It further identifies terrorist groups that have caused the highest number of casualties world-wide over the study period.

While one might believe that Europe and North America are more vulnerable to terrorism than other continents, Chapter 4 brings empirical evidence that disproves this belief. It compares both the number and intensity of terrorist attacks for each continent and its evolution over time, from 2002 to 2017. Furthermore, it highlights the most affected countries and cities by terrorism.

Has terrorism increased over time, and if so, how long did it last and where has the increase been observed? Chapter 5 reveals trends in monthly terrorist events and associated deaths from 2002 to 2017 that can be identified at various spatial scales: worldwide, continent-level, country level (with a focus on the four most affected countries by terrorism: Iraq, Pakistan, India, and Afghanistan), and at city-level with an example applied to most targeted city in the world: Baghdad, Iraq.

While one might believe that terrorist events occur "randomly", evidence indicates that terrorist events are not the result of dice rolling and tend to be clustered at various spatial scales. Chapter 6 highlights impact of spatial autocorrelation and the role of the scale in the analysis of spatial clusters of terrorism. Furthermore, using the city of Baghdad as an illustration, it shows the limitations of the Global Terrorism Database (GTD) with regard to spatial accuracy and compares the results of a simulation of potential locations of terrorist attacks within the city of Baghdad using various point processes. This chapter provides evidence that the clustering process observed may be associated with the selection of specific targets by terrorists, which explains the observed clusters of terrorist attacks.

Chapter 7 focuses on the spread (also called "diffusion") of terrorism, which appears within areas facing an abnormally high risk ("hotspots") of terrorist attacks. Furthermore, it describes three common approaches to consider spatial processes: lattice, geostatistical, and point process model. Based on the analysis of point patterns, the chapter provides evidence that the location and size of hotspots can vary over time, as exemplified in the analysis of monthly terrorist attacks perpetrated by the Islamic State of Iraq and al Shame (also called Islamic State or ISIS) in Iraq in 2017. The analysis highlights the fall of ISIS in Iraq, illustrated by shrinkage (dissipation) of terrorism's hotspots observed over the year 2017.

Chapter 8 describes the role and limitations of stochastic models to predict natural and social phenomena, including terrorism. It distinguishes deterministic from stochastic models, and highlights the potential of the latter as a suitable framework to predict terrorism since it may rigorously account for uncertainty. It introduces the reader to algorithmic models which can predict terrorism at fine spatial and temporal scale. It describes a machine learning algorithm approach used to predict terrorism a week ahead at fine spatial scale in Iraq, Afghanistan and Pakistan.

Chapter 9 attempts to distinguish what is currently known from what is currently not known in the field of terrorism. Furthermore, it suggests possible future areas of research that could bring additional knowledge and policy-relevant guidance, mainly through the development of innovative statistical methods, increase of computational power, and access to more comprehensive datasets.

Debunking Seven Terrorism Myths Using Statistics provides a compact resource on major findings in terrorism that have been recently gained through the analysis of global databases on terrorism. It offers a rigorous framework along a few statistical techniques required to critically analyze and interpret terrorism data. Furthermore, it reminds the reader about the ephemerality of terrorism knowledge and uncertainty of the very nature of this complex phenomenon.

CHAPTER 2

Myth No 1: We Know Terrorism When We See It

> Terror consists mostly of useless cruelties perpetrated by frightened people in order to reassure themselves.
>
> Friedrich Engels

2.1 Introduction: the necessity to interpret terrorism data with caution

What is terrorism? What can we learn from terrorism data? What is hidden from the data? What can make the analysis of terrorism data misleading? Through various case studies, this chapter illustrates why one should interpret the results of any analysis of terrorism data with caution.

We will discuss major issues that affect the analysis of terrorism data. We will expose the inherent ambiguity of the concept of terrorism and explain its impact on its understanding. We will illustrate discrepancies between spatial patterns of terrorist events based on data gathered from providers that use different methodologies to gather terrorism data.

Terrorism is often classified into classes to highlight or compare specific aspects of it, such as its lethality or the type of targets. However, the choices made to classify data are hardly without consequences. We will illustrate the impact on the interpretation of the results of analysis using different approaches to classify terrorism by target types, actors, and purpose.

2.2 No consensus on the definition

Scholars in terrorism are unanimous. Terrorism cannot be reduced to a single explanation. Terrorism is plural in every sense; its essence is controversial. Consensus on the definition of terrorism has not been reached neither in academia, nor among governments.

In the UK, the Terrorism Act 2000 defines terrorism as: "the use or threat of action where (a) the action falls within subsection (2), (b) the use or threat is designed to influence the government or an international governmental organization or to intimidate the public or a section of the public, and (c) the use or threat is made for the purpose of advancing a political, religious, racial or ideological cause".

Some actions are considered as terrorist act (subsection 2): if they "(a) involve serious violence against a person, (b) involve serious damage to property, (c) endanger a person's life, other than that of the person committing the action, (d) creates a serious risk to the health or safety of the public, or (e) is designed seriously to interfere with or seriously to disrupt an electronic system" (Terrorism Act 2010 (c. 11, s. 1)).

The British government definition of terrorism includes cyber-terrorism (see 2e), which is typically excluded from terrorism databases such as the *Global Terrorism Database* (GTD) [56]. At international level, there is currently no agreement on a legal definition of terrorism that would be applicable for all or at least most countries. The definition of terrorism suggested by the United Nations Security Council in 2004 remained *non-binding* towards Member States, which signifies that it has no legal weight [129].

In order to find a common denominator among academic definitions of terrorism, one can count the most frequent concepts used to define terrorism. Based on a large pool of academic definitions, the most cited concepts in order of frequency are: "violence-force", "political", "fear-terror", and "threat" [127, 144]. Finding the most frequent words is a useful exploratory tool but does not suffice to account for the diversity of views on terrorism.

Several equally valuable views on terrorism may coexist and some degree of subjectivity in the interpretation of the concept of terrorism cannot be avoided. Furthermore, any single definition of terrorism is ineluctably incomplete and subject to debate [47, 63, 8, 53].

Below, we attempt to define terrorism based on four characteristics:

1. *Political aim*: there is a political intention behind acts of terrorism;

2. *Fear*: terrorists aim at generating fear;

3. *Publicity*: terrorists look for publicity;

4. *Civilian target*: terrorists may deliberately target civilians.

This list remains ineluctably incomplete and unable to capture all possible interpretations of the concept of terrorism. Nevertheless, it has the advantage of identifying key characteristics of terrorism inline with the definitions adopted by most scholars and providers of terrorism data. Furthermore, it allows differentiating terrorism from related forms of political violence, such as war, riots, or robbery to name but three. The shortbox provides further details on the legal ambiguity of terrorism and comment a definition from a scholar.

Terrorism is commonly assumed to be politically driven unlike other crimes, which find their motivation in different reasons. For example, crime of passion is mainly driven by personal rather than political motives (first criterion: *Political aim*). Furthermore, terrorism often aims at generating fear *per se* (second criterion: *Fear*), as suggested by its Latin root *terrere*, which literally means "to frighten". Other forms of crime, such as burglary for example, do not primarily intend to create fear. Rather, burglary is mainly driven by economic motives. Even though victims of burglary may also be frightened, fear is usually an indirect consequence of burglary.

The actions perpetrated by terrorist groups often intend to provoke a disproportionate reaction from the government, which, in turn, may provoke anger among the population targeted by repressive government actions. Consequently, the discontent of the population with the authorities may increase and be expressed through further support for the terrorists (third criterion: *Publicity*).

Likewise, a deliberated attempt at targeting civilian population (fourth criterion: *Civilian target*) has been a tactic of predilection used by numerous terrorist groups, including Al Qaeda. This tactic is peculiar for terrorism as it is forbidden by international laws. The Agreements of 1949 of the Geneva Conventions state that civilians are not to be subject to attack within the context of war [66, 27, 128, 63]. However, these conventions are often violated, and therefore, deliberate acts of violence on civilians cannot be totally excluded during war.

Terrorism can be conducted and perpetrated by *state* and *non-state* actors. A terrorist attack executed by a government refers to state terrorism. The concept of terrorism originally refers to 18th century terrorism perpetrated by governments, which was defined as "violent acts of governments designed to ensure popular submission" [16, pp. vii]. Friedrich Engels' quote on terror that introduces this chapter refers to state terrorism perpetrated during the Reign of Terror, a violent period during the French Revolution in which the French government was carrying out public executions that targeted enemies of the revolution.

For example, the Stalin, Hitler, and Pol Pot totalitarian regimes committed mass violence, including state terrorism [53]. One may also consider the atomic bombs dropped on Hiroshima and Nagasaki in August 1945 as state terrorism. Other US military operations can be viewed as state terrorism: the "Operation Just Cause" which killed several thousands of people in Panama in 1989, or the use of terror by US military in Nicaragua; the latter has been condemned by the International Court of Justice in 1986 [68, 71] quoted in [17, pp. 16-18].

Other cases of state terrorism include Russia's attacks in Chechnya in the 1990s, the Israeli invasion of Lebanon in 1982, the use of terror in the 1990s by Turkey in the Kurdish areas, crimes committed by the Colombian army against trade unionist and journalists during the same period [17, pp. 52, 62-67], or the use of terror during the military regime *Proceso de Reorganización Nacional* in Argentina from 1976 to 1983 [110].

States may also commit terrorist attacks through the intermediary of one or several non-state actor(s). In this scenario, states do not commit the attack but provide support (e.g. money, weapons, military advice and training) to non-state groups. This commonly refers to *state-sponsored* terrorism. The distinction between state and non-state terrorism is often difficult to make since information on the exact role of states in terrorist operations is usually not disclosed to public.

In line with most practitioners and terrorism database providers [38, 20, 45, 46, 92, 12, 79], this book considers almost exclusively *non-state* terrorism. For the sake of conciseness, we will usually omit the *non-state* attribute and equate non-state terrorism with terrorism when not stated otherwise. However, in Section 2.5 we compare the location and intensity of violent events perpetrated by state actors with those carried out by non-state actors and will therefore consider both state and non-state terrorism.

2.3 Discrepancies among databases

There are several sources on terrorism that provide the spatial coordinates of terrorist events—data with spatial coordinates (e.g. longitude and latitude) are also called geolocalized data—across the globe: the *Global Terrorism Database* (GTD) [56], the *RAND Database of Worldwide Terrorism Incidents* (RDWTI) [118], and the *Global Database of Events, Language, and Tone* (GDELT) [84]. Although other databases, such as the *International Terrorism: Attributes of Terrorist Events* (ITERATE) have been extensively referred to in terrorism research, they do not geolocalize events (neither the name of the targeted city nor geographic coordinates are provided).

Terrorism data are problematic for several reasons. Terrorism databases usually gather information from media sources but the media do not systematically provide reports on terroristic events. One source of (reporting) bias comes from the fact that events reported by media are not missing at random [132]. Instead, terrorist events tend to be less reported in countries where freedom of press is reduced [38, 37]. In other words, one tends to underestimate the true number of terrorist events that occur in countries with restrictions on press freedom.

Furthermore, each terrorism database adopts a specific view on terrorism. The classification of acts as terrorist events depends on the underlying definition of the concept of terrorism and the process applied to gather, code, and classify data. This process can be fully or partially automatized using, e.g., machine learning algorithms used as an initial filter to eliminate events that are unlikely to be terrorist events. The process used to classify terrorism data is unique to each data provider.

One way to illustrate discrepancies among databases consists of comparing the location of terrorist events gathered from multiple terrorism databases in the same geographical area and time period, to allow for a sensible comparison. Figure 2.1 shows the locations of terrorist attacks that occurred in the world from 2002 to 2009, according to data provided by three databases: GTD, RDWTI, and GDELT.

The study period used to compare the location of events does not extend beyond 2009, which corresponds to the latest data available in RDWTI. We started the comparison in 2002, which is often used as reference year, because it marks a new "era" of global terrorism, the one starting the year after the 9/11 events and bearing profound consequences on society.

Figure 2.1 illustrates discrepancies among the locations of terrorist events gathered from different databases: (*red triangle*: RDWTI, [118], *black circle*: GTD [56], and *green cross*: GDELT [84]). Had the data on terrorist events been coded, classified and gathered from the same media reports using a common methodology (including an identical definition of terrorism) shared among the databases, GTD, RDWTI, and GDELT events should overlap.

. Instead, a complete overlap between locations is not observed. Numerous events do not superimpose on the map. The differences observed in the location of the events indicate discrepancies among the databases in the geolocation of terrorist events. Further divergences, such as the number of deaths, wounded, type of weapon used, or the name of the perpetrators of the attacks are also common when different databases are compared.

Recent approaches have been developed to improve the empirical measurement of conflict events based on an automated procedure that combines multiple conflict and terrorism datasets. For further interest, we refer the reader to the paper *Integrating conflict event data* from Donnay et al. published in 2019 in the *Journal of Conflict Resolution* [35] along with their R package `meltt` "Matching Event Data by Location, Time and Type" for applications.

2.4   Side effects of distinguishing targets

Terrorism can occur in both war and peace contexts. Civil war—an armed conflict between government and civilian forces over a specific period that cause a large number of deadly casualties—may provide opportunities for terrorist groups to attack government forces [125, p. 174]. In a wartime context, one may analyze terrorist attacks perpetrated against the government, and hence, select data that include only attacks that target non-civilians governmental actors.

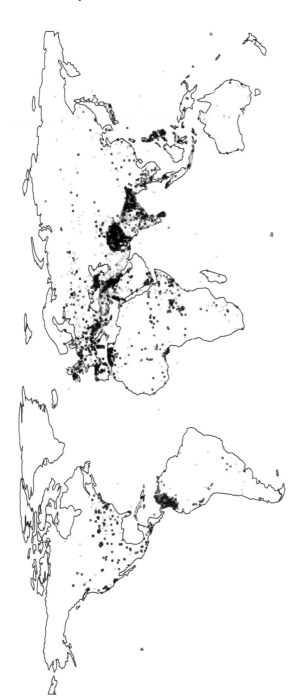

Figure 2.1: Mapping worldwide terrorist events from 2002 to 2009, according to three database providers. Provider n°1: *red triangle* RAND Database of Worldwide Terrorism Incidents (RDWTI) [118], provider n°2: *black circle* Global Terrorism Database (GTD) [56], and provider n°3: *green cross* Global Database of Events, Language, and Tone (GDELT) [84]. Image obtained from [113].

In other situations, terrorist groups may focus their attacks on civilian targets to coerce the opposing government into changing its policy or to change the population's behavior [108]. The extreme distress caused by suicide attacks on civilians to society may lead the government to capitulate to the requests of the terrorists [108, p. 346]. In this context, one may exclude attacks that target non-civilians, such as military forces or police. How can a definitional restriction based on the type of target affect the observed spatial patterns of terrorist events? How data selection might affect the interpretation of the resulting pattern and its impact on policy recommendations?

Using the GTD dataset [56], we map the location and the number of terrorist attacks perpetrated against civilians (Figure 2.2) and non-civilians (Figure 2.3) in northern Africa from 2002 to 2017. For illustrative purposes, we focus on northern African countries, which include Algeria, Egypt, Libya, Morocco, South Sudan, Sudan, and Tunisia. To ease comparison, we aggregate the total

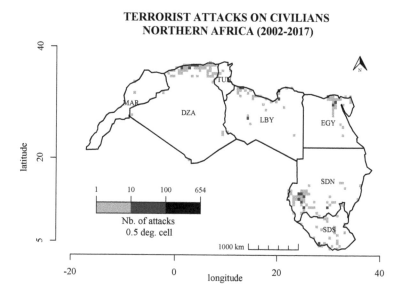

Figure 2.2: Terrorist attacks perpetrated against civilians aggregated within 0.5 degree grid-cells from 2002 to 2017 in North Africa: Algeria (DZA), Egypt (EGY), Libya (LBY), Morocco (MAR), South Sudan (SDS), Sudan (SDN), and Tunisia (TUN). Data source: GTD [56].

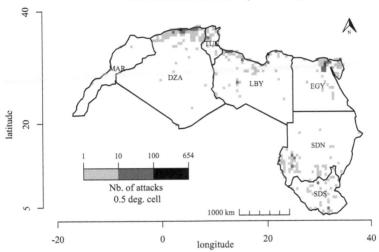

Figure 2.3: Terrorist attacks perpetrated against non-civilians aggregated within 0.5 degree grid-cells from 2002 to 2017 in North Africa: Algeria (DZA), Egypt (EGY), Libya (LBY), Morocco (MAR), South Sudan (SDS), Sudan (SDN), and Tunisia (TUN). Data source: GTD [56].

number of attacks into 0.5 degree regular grid-cells, which correspond to cells of about 55 km of resolution at the Equator.

In Figure 2.2, civilian targets include: attacks on airport and aircraft, educational institution (e.g. teaching staff or school buses), private citizens and property (e.g. public markets), religious figures and institutions (e.g. mosques or churches).

We compare the number of deadly casualties provoked by attacks perpetrated against civilians (*white dots*) and non-civilians (*black dots*) (Figure 2.5). From 2002 to 2017, the total number of deaths was relatively similar between Algeria and Libya. Algeria counted 1825 deaths while Libya mourned 1621 deaths from terrorism. However, the proportion of civilians killed by terrorism in Algeria was almost twice the values observed in Libya. In Libya, one fifth of all deadly terrorist attacks targeted civilians. In Algeria, terrorist groups aimed their attacks on a larger proportion of noncombatants. In the same period, more than one third of deadly attacks were civilians. When distinguishing terrorism by target type, one should keep in mind that the total number of deaths and the proportion of civilian deaths metrics may tell a different story.

The number of attacks and the location of the attacks is also affected by restricting the definition of terrorism by civilian and non-civilian targets, as illustrated in Figures 2.2, 2.3, and 2.4. From 2002 to 2017, Libya counted the highest number of terrorist attacks among the investigated countries in North Africa. From a total of 1323 attacks less than one quarter targeted civilians (Figure 2.4). During the same period, Sudan recorded just above one fifth of the total number of attacks that occurred in Libya (405 attacks). However, the proportion of attacks on civilians remains high. More than a half of the attacks targeted civilians.

Did Morocco encounter more deaths due to terrorism than Tunisia? This question may appear simple, however it is in fact difficult to answer without further conceptual clarification. In Tunisia, terrorist attacks killed 6 civilians and 234 non-civilians. If, for conceptual reasons, one excludes terrorist attacks on non-civilians from the dataset, Tunisia exhibits far less death casualties (6 civilian deaths) than Morocco (25 civilian deaths). However, when the death toll of both civilians and non-civilians is added, Tunisia

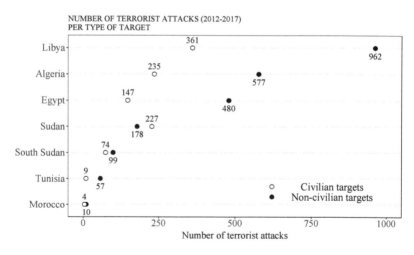

Figure 2.4: Total number of terrorist attacks perpetrated against civilians (*white dot*) and non-civilians (*black dot*) from 2002 to 2017. The investigated countries are: Libya, Algeria, Egypt, Sudan, South Sudan , Tunisia, Morocco. Data source: GTD [56].

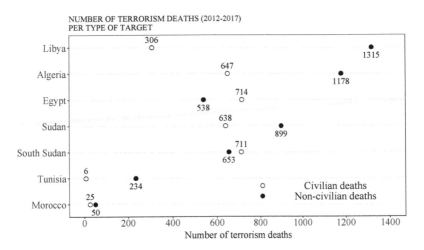

Figure 2.5: Total number of deaths due to terrorist attacks perpetrated against civilians (*white dot*) and non-civilians (*black dot*) from 2002 to 2017. The investigated countries are: Libya, Algeria, Egypt, Sudan, South Sudan , Tunisia, Morocco. Data source: GTD [56].

counts more than three times the number of death casualties than those that occurred in Morocco.

In order to analyze several facets of terrorism, one may require to subset data and exclude events from specific classes or categories. However, the exclusion of data may affect the results of analysis of terrorism data. This may lead to considerable changes in the observed spatial location, number of attacks, and number of deaths.

Hence, interpreting the risk of terrorism and developing or implementing counterterrorism policy from the analysis of subsets of terroism data requires particular care. To mitigate the risk of misinterpretation, it is crucial to clearly state the source of the terrorism data used in the analysis and mention any filters that have been applied to subset the database. When possible, comparing results with different sources of data can prove to be helpful.

2.5 State repression and non-state terrorism: insight from the Democratic republic of Congo

As mentioned in the introductory chapter, state actors may perpetrate violent repressive measures against non-governmental targets, including acts that lead to death casualties, which refer to as lethal events. The Uppsala Conflict Data Program (UCDP) has made available the *Georeferenced Event Dataset* (GED), which provides data on lethal events perpetrated by governmental forces against non-governmental actors until 2016 (this might be updated in the future) [137].

From 2002 to 2016, the Democratic Republic of Congo (DRC) encountered 526 lethal state attacks and 210 lethal non-state terrorist events. Figure 2.6 illustrates the location of lethal state terrorist attacks (*left*) [137] and non-state lethal terrorism (*right*) [56] in the DRC from 2002 to 2016. Some parts of the country (center-west, north, north-east, and south-east) are highly targeted by both state and non-state lethal events.

The number of GED lethal attacks perpetrated by state actors (527) is more than twice the number of GTD lethal attacks perpetrated by non-state actors (210). In the DRC, lethal repressive measures carried out by state actors seem to overtake those perpetrated by non-state actors. One should however interpret this result with caution. The data gathering process and the conceptual definitions applied to select events may substantively differ among GED (state actors) and GTD (non-state actors). Inconsistency in the data processing method and divergences between the definitions of terrorism used by GED and GTD cannot be excluded, and therefore, make comparisons challenging.

We have shown several differences between databases on terrorism that can be visualized, such as the location of events and their intensity (e.g. number of attacks per given area). A comparison between databases is a useful exploratory step and the information resulting from this comparative work can help select the database for further analysis and identify the limitations of the data.

2.6 Political and non-political terrorism: lessons learned from Pakistan

How can we distinguish terrorism from other forms of violence? If the main aim of a reported violent event is political, then it is often classified as a terrorist event (see Section 2.2). A counterexample would be crime of passion, which is usually driven by personal motives rather than political ambitions.

Can we claim that terrorism is necessarily political? It depends. Some authors, however, argue that some groups may use terrorist tactics in order to achieve an economic, religious, or social goal

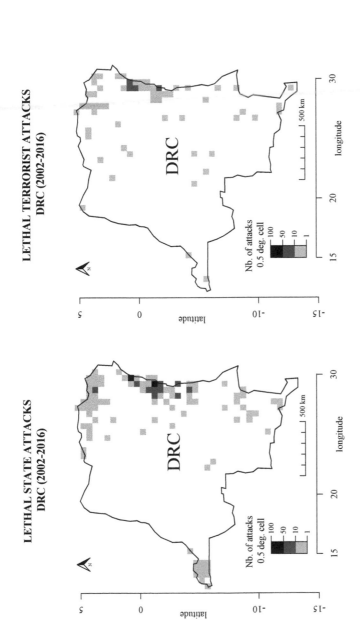

Figure 2.6: Lethal attacks perpetrated by governmental forces (*left*) and non-state terrorist groups (*right*) in the Democratic Republic of Congo (DRC) from 2002 to 2016. Total attacks aggregated into 0.5 degree grid-cell. Data source: state lethal events: UCDP [137] and non-state terrorist attacks: GTD [56].

instead of being driven by political motives. For example, terrorist groups might use kidnapping tactics to obtain ransom. In this case, one may consider this type of act as a non-political terrorist event.

GTD [56] proves very useful to explore the differences between political from non-political terrorist events. For the case of illustration, we focus on events that occurred in Pakistan from 2002 to 2017. Figure 2.7 shows the number of political (*left*) and non-political (*right*) terrorist attacks that occurred within 0.5-degree grid-cells. The number of political terrorist attacks is larger and affects larger areas compared to non-political terrorist events.

Pakistan hosts various notorious home-based and foreign-based terrorist groups. Several branches of Al-Qaeda operated in Pakistan in 2017 [19]. From 2002 to 2017, 9,790 terrorist attacks in Pakistan were politically motivated, while 146 attacks pursued other goals. In this case, a study that would exclude terrorist events perpetrated for political motives would substantially underestimate the true number of terrorist events.

Further analysis could be carried out to investigate a possible link between the type of attack and the perpetrator. One might reasonably believe that groups that, e.g., claim territory are more likely to carry out terrorist attacks for political motives. Moreover, terrorism can be distinguished using different classification (e.g. by size of the group, by country of origin of the perpetrator, etc.).

2.7 Conclusion

This chapter has pointed out some differences in terrorism data that may occur due to: (a) collection methodologies, data sources, scope, and measurement errors that may vary according to each database (e.g. GDELT, GTD, RDWTI); (b) a split of terrorism data into categories, such as: civilian versus non-civilian targets, political versus non-political events, and (iii) state versus non-state terrorism.

Through the use of the appropriate techniques, variations among the databases can be explained. Comparing the results from different databases remains a crucial part of any exploratory analysis aiming at providing transparency in the description of terrorism. Dividing terrorism data into categories may highlight different facets. However, this operation can lead to important changes in the observed pattern, and hence, its interpretation.

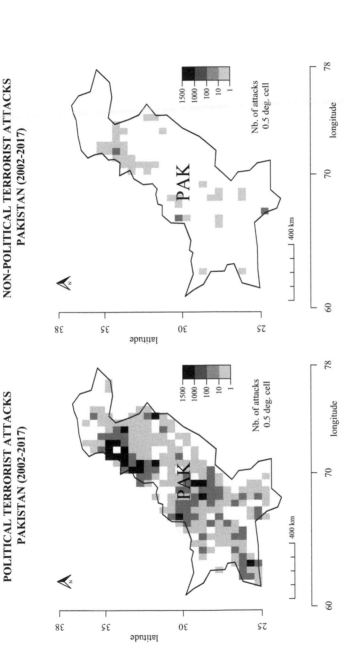

Figure 2.7: Political (*left*) and non-political (*right*) terrorist events perpetrated by non-state actors in Pakistan from 2002 to 2017. Total attacks aggregated into 0.5 degree grid-cell. Data source: GTD [56].

It is worth remembering that no consensus about the definition of terrorism has been reached, which considerably restricts the generalizability of findings made in the field of terrorism research. Furthermore, the data gathering process may vary according to each database provider. Most databases on terrorism exclude de facto some terrorist events such as state terrorism, planned but failed attacks, and/or canceled attacks [78].

When we visualize terrorist data we only "see" terrorism as defined and interpreted by the database provider and restricted by potential additional filters that might have been applied to classify data. Visualization of data is merely a mirror that reflects a particular view of terrorism, that may or may not be shared by other database providers or scholars.

We emphasize that the inherent subjectivity in the definition of terrorism does not imply that any view on it is valid; rather, it calls for vigilance in the interpretation of the analysis terrorism data.

With regard to the inherent limitations of data, reliable research work should: (1) provide transparency with regard to the source of data, clearly formulate the definition(s) used and acknowledge potential problems inherent to the data; (2) consider how dividing terrorism data into subsets impacts the validity of the results; (3) when possible, compare different databases or subsets of data from the same database and acknowledge discrepancies in the methods applied to gather data; (4) offer nuanced interpretation when results are communicated.

About the definition of terrorism

"There is no universal legal definition approved by the General Assembly of the United Nations (the one proposed by the Security Council in Res. 1566 (2004) is non-binding, lacking legal authority in international law)" [129]. The concept of terrorism is intrinsically ambiguous and still being debated today [8, 53].

Hoffman attempted to formulate a concise definition of terrorism: "[...] the deliberate creation and exploitation of fear through violence or the threat of violence in the pursuit of political change" [63, p. 40]. Despite its remarkable conciseness, Hoffman acknowledged the incompleteness of his definition: subjectivity is almost inevitable when groups or individuals are defined as terrorists [63, p. 23]. Inevitably, the definition of terrorism implies a judgment [105, p. 91].

Glossary

Lethal terrorism: terrorist attacks that lead to one or more deaths.

Non-lethal terrorism: terrorist attacks that do not cause death casualty but may result in injuries.

Non-state terrorism: terrorism perpetrated by non-state actors (sponsored or not by state actors) against civilian or non-civilian targets. Example: The attacks perpetrated by Al Qaeda (9/11), which killed about 2,996 people in the US on 11 September 2001.

State terrorism: terrorism perpetrated by a state actor against its own citizens or against targets outside its national boundaries. Example: US nuclear attacks on Hiroshima and Nagasaki in August 1945.

FURTHER READING

Wilkinson, P. (1979). Terrorism and the Liberal State (2nd ed.). *The MacMillan Press Ltd, Hong Kong, China.*

Richardson, L. (2006). What Terrorists Want (1st ed.). *John Murray, London, UK.*

Hoffman, B. (2006). Inside Terrorism (Rev. and expanded ed.). *Columbia University Press, New York, NY, USA.*

English, R. (2010). Terrorism: How to respond. *Oxford University Press, Oxford, UK.*

CHAPTER 3

Myth No 2: Terrorism Only Aims At Killing Civilians

> What if the enemy advances
> to the attack with a large,
> well-organized army? Seize
> something they value highly,
> then they will listen to you.
>
> Sun Tzu

3.1 Introduction: a note of caution on the validity of the analysis of terrorism data

In this section and the following ones, we will analyze several aspects of terrorism based on GTD, which is currently the world's most extensive unclassified database on terrorism [80]. As mentioned earlier in the book, all terrorist databases suffer from various issues that affect the reliability of the data and GTD is no exception. One should keep in mind that the results and their interpretation described further in the text only apply within a relatively restricted scope, constrained by the definition used by the provider and limited by inherent issues in the data collection and classification processes.

Our working definition of terrorism considers acts as terrorism if at least two of the following three criteria are satisfied: *(i) "The act must be aimed at attaining a political, economic, religious, or social goal"; (ii) "there must be evidence of an intention to coerce, intimidate, or convey some other message to a larger audience (or audiences) than the immediate victims"; (iii) "the action must be outside the context of legitimate warfare activities" [56].*

Furthermore, we consider only events that are spatially geolocated to make analysis at fine spatial scale. Therefore, events with missing values for longitude or latitude are excluded. The trade-off here is that we underestimate the true number of attacks while increasing spatial accuracy of our analysis. Furthermore, we focus our analysis on the lethality of terrorism, or in other words, the propensity of terrorism to kill individuals. Therefore, we keep events that provide indication on the number of deaths. Any further reduction of the dataset made in the analysis carried out in this chapter is indicated in the corresponding text.

3.2 Half of the terrorist attacks do not kill

While it is commonly accepted that terrorism often aims at spreading fear and terror beyond its immediate victims, it would be, however, misleading to believe that terrorism is only about killing civilians. Following the Geneva Conventions of 1949 (Article 50 of Additional Protocol I adopted in 1977), we understand civilians as people who are not combatants or members of the military [67].

According to GTD [56], from 2002 to 2017, about half the 75,906 terrorist attacks perpetrated in the world caused one or more deaths; the other half may have caused injuries but did not engender any death. Over this 16-year period, terrorism killed more than 190,000 individuals in the world.

The *median* number of deaths per terrorist attack is one. The *average* number of deaths per attack is 2.5 (see the shortbox for further details on the interpretation of the mean and median in this case study). On a total of 75,906 attacks, 48% of them (36,155) did not kill and 52% of the attacks (39,751) generated one or more death casualties. This means that from 2002 to 2017, there were about the same number of lethal attacks as non-lethal attacks worldwide.

Close inspection reveals that the proportion of lethal versus non-lethal attack varies across continents. Figure 3.1 shows that in Asia and Africa, the proportion of lethal attacks is higher than non-lethal attacks. Note that the x-axis is in $\log_{10}$ scale, which means that each successive tick indicates a value 10 times higher than the previous one.

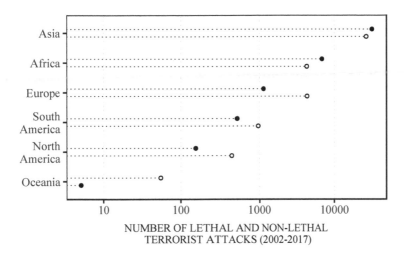

Figure 3.1: Number of lethal (*black dot*) and non-lethal (*white dot*) terrorist attacks perpetrated from 2002 to 2017 in Asia, Africa, Europe, South America, North America, and Oceania. *Ticks placed on Log_{10} scale.* Data source: GTD [56].

In contrast, Europe and the Americas exhibit a higher proportion of non-lethal attacks. While the proportion of lethal attacks versus non-lethal attacks is about equal worldwide, an analysis at finer spatial scale reveals heterogeneity in the lethality of terrorism among continents. The proportion of lethal attacks in Asia (54%) and Africa (62%) is far above those observed in Europe (21%) and South America (35%) and North America (26%).

However, one should take these results with a caveat. As mentioned in Section 2.3, selective bias—more events tend to be reported in countries with higher freedom of press—cannot be excluded. Both the reported number and proportion of lethal attacks may deviate from true values.

3.3 Measuring and interpreting terrorism casualty is affected by data classification

The magnitude or impact of a terrorist event is often measured as the number of casualties it has engendered, and it is common to

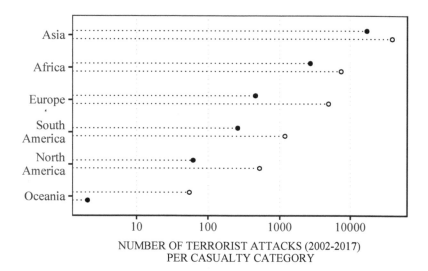

Figure 3.2: Number of low casualty (*white dots*: < 5 killed or wounded) and high casualty (*black dots*: >= 5 killed or wounded) terrorist attacks perpetrated from 2002 to 2017 in Africa, Asia, Europe, North America, Oceania, and South America. *Ticks placed on Log_{10} scale.* Data source: GTD [56]. Terrorist attacks that do not provide information on the number of wounded or killed (2,591 events) are excluded.

transform values into classes to summarize the data. While such classification can be useful, it should be done with caution.

For example, Stanton [135] suggests classifying terrorism into two classes: high casualty and low casualty events. How should one define the size and limits of the classes? How does the approach used to define classes of casualty affect the results? Below, we propose several ways of classifying data into classes based on different thresholds and variables associated with casualties. We point out the challenges in interpreting results on the magnitude of terrorism from data rearranged into classes.

In Figure 3.2, we distinguish two classes. Low casualty events (*white dots*) include attacks with less than five killed or wounded, while high casualty attacks (*black dots*) encompass events that lead to 5 or more killed or wounded. In this approach, killed and wounded have the same weight in the summation. For example, an attack that kills 3 people and wounds one person is classified as low casualty (total of 4 casualties) while an attack that kills 2 people and wounds 3 individuals (total of 5 casualties) is considered a high casualty terrorist event.

In the GTD dataset, the maximum number of deaths for a single terrorist attack observed from 2002 to 2017 is 1,570 [56]. This corresponds to an attack perpetrated on June 12, 2014, by the Islamic State (ISIS) in the city of Tikrit, Iraq. Despite their devastating effects on society, mass-casualty attacks—attacks that kill a large number of people—are rare.

Using the aforementioned classification approach, one observes about 2.5 times more low casualty events (52,890 events) than high casualty events (20,425 events) on a global scale. Consistently, Figure 3.2 shows that all continents exhibit a higher proportion of low casualty events (*white dots*) compared to high casualty (*black dots*) events.

How changes in classes intervals would affect these results? As an alternative classification approach, we define three categories, as illustrated in Figure 3.3: low-casualty attacks (*white dots*: 22,906 events, which count 0 death or wounded); *gray dots*: intermediate casualty attacks (41,146 events, which count between 1 and 10 deaths or wounded); *black dots*: high casualty attacks (9,263 events, which count more than 10 deaths or wounded).

This new classification offers further nuances in the interpretation of the previous results. Asia and Africa tend to encounter a higher proportion of intermediary casualty attacks (*gray dots*) compared to low-casualty (*white dots*) and high casualty (*black dots*) events. However, Europe exhibits a higher proportion of low-casualty (*white dots*) compared to intermediary casualty attacks (*gray dots*) and high casualty (*black dots*) events.

Should wounded be included as casualty events? Should the number of wounded be weighted equally as deaths when summed into a casualty index and attributed to a specific class? Should we include other factors, such as economic loss when casualty is used to estimate the magnitude of terrorist events? The definition of levels of casualty is inherently subjective: there is no right or wrong classification of terrorism casualty. However, one should be aware that the type of events considered (wounded, deaths, etc.) and the number and lengths of the classes may affect the observed patterns, and hence, the interpretation of the results.

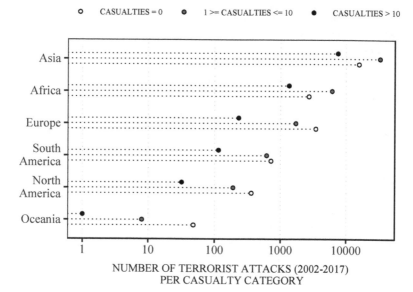

Figure 3.3: Number of low casualty (*white dots*: 0 killed or wounded); intermediary casualty (*gray dots*: 1 − 10 killed or wounded; high casualty (*black dots*: > 10 killed or wounded) terrorist attacks perpetrated from 2002 to 2017 (75,906 events) in Africa, Asia, Europe, North America, Oceania, and South America. *Ticks placed on Log_{10} scale.* Data source: GTD [56]. Terrorist attacks that do not provide information on the number of wounded or killed (2,591 events) are excluded.

3.4   Witnessing levels of terrorism violence: a focus on the Islamic State

The number of casualties caused by terrorism varies among the perpetrators of the attacks. For terrorist groups that mainly depend on large civilian constituency, inflicting a high number of civilian casualties could be costly and jeopardize the survival of the group. This could lead to a loss of credibility of the terrorist group and reduction of support from its own civilian constituency. However, the degree of dependence of terrorist groups towards their constituency may vary. It is expected that terrorist groups who require less support from civilians are less likely to perpetrate high casualty attacks [135].

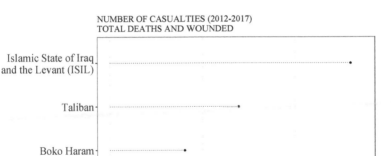

NUMBER OF CASUALTIES (2012-2017)
TOTAL DEATHS AND WOUNDED

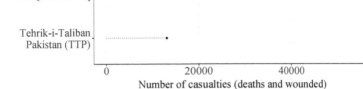

Number of casualties (deaths and wounded)

Figure 3.4: Terrorist groups that shows the five highest total number of casualties (killed and wounded) worldwide from 2002 to 2017 (75,906 events) are highlighted. Data source: GTD [56].

While most terrorist groups do not seek to provoke a high number of casualties, a few terrorist groups behave radically differently. Some groups may inflict large human losses. Figure 3.4 ranks the five terrorist groups that generated the highest number of casualties (killed and wounded) from 2002 to 2017. Note that there is high uncertainty with regard to this ranking. These statistics exclude events whose perpetrator is not known. In the GTD database, this corresponds to a total of 197,269 casualties. Should the perpetrator be known, this might affect the results of this analysis.

The terrorist groups (known perpetrators) with the highest number of casualties are (in order): The Islamic State of Iraq and the Levant (ISIL), Taliban, Boko Haram, Al-Qaeda in Iraq, and Tehrik-i-Taliban Pakistan (TTP). Note that Al-Qaeda in Iraq is the precursor of ISIL (see further information on these terrorist groups in the Glossary).

Most casualties (name of the main perpetrators in parentheses) are related to terrorist attacks located in Iraq (ISIL and Al-Qaeda in Iraq), Afghanistan (Taliban), Nigeria (Boko Haram), and Pakistan (TTP). From 2002 to 2017, ISIL has been by far the most deadly terrorist groups with almost twice as many casualties as the Taliban.

3.5 Conclusion: terrorism does not ineluctably equate with the death of civilians

In this chapter, we have discussed approaches that aim at estimating the magnitude of terrorist event by focusing on casualties, which include the number of killed and wounded people. We showed that different classification methods to capture lethality and the number of casualties directly impact on the results and their interpretation.

The definition of the classes, which includes both the number of classes and their intervals, used to distinguish the magnitude of terrorist events with regard to the casualties has an important impact on the interpretation of the results. Furthermore, other threats to the validity of these results should be considered. Data on fatality count may be more accurate than data on wounded. The latter might vary among countries due to the use of diverse criteria used to declare people as wounded.

We have also mentioned that most terrorist groups do not cause large casualty attacks. Few terrorist groups, mainly operating in Iraq, Afghanistan, Nigeria, and Pakistan, are exceptions and may generate a large number of casualties. We attempted to identify terrorist groups who commit the highest number of casualties (killed and wounded) from 2002 to 2017.

Using GTD, we observed that most casualties are perpetrated by unknown terrorist groups. This indicates that uncertainty remains with regard to the identification of the perpetrator of terrorist attacks, and hence, the identification of the most lethal groups. A higher proportion of labeled attacks would be required to link the perpetrators with the observed terrorist attacks worldwide.

For the sake of transparency, we encourage analysts rearranging terrorist events into classes to compare results using different classification approaches and to discuss the results of their analysis. The methodology used to classify the level of violence of terrorist events may substantially affect the results and their interpretation, and hence, the validity of the conclusions.

From a scientific perspective, identifying differences in results that arise from changes in the classification of events is crucial. It can be used to assess the robustness of the models to one type of model specification change. Using specific thresholds provided by experts to determine the classes may also reveal subtle mechanisms that lead to generate different types of attacks.

Impact of the choice of a central tendency indicator to represent trend in terrorism

There are various ways to represent a central tendency in the data, and each indicator has specific properties. The average is sensitive to extreme values in contrast to the median; the latter is therefore often favored for its robustness. How may the choice of the central tendency indicator impact our understanding of terrorism? Assume that the following list of numbers represents the number of deaths of seven terrorist attacks that occurred in a given location: $\{0, 0, 6, 0, 1, 1, 1\}$.

If one ranks these numbers from the lowest to the highest value $\{0, 0, 0, \mathbf{1}, 1, 1, 6\}$, the median corresponds to 1, which is the central (4th) element of the list (in bold). The average, instead, is computed as the sum of each element divided by the number of elements ($\frac{0+0+0+1+1+1+6}{7} = 9/7 \approx 1.3$).

The analysis of the difference between the median and the average is informative. If the average is higher than the median, it suggests the presence of one or few terrorist attacks that count a number of deaths substantially higher than the median. For example, a mass-casualty attack—an attack that causes an extremely large number of deaths—can substantially increase the average number of deaths without necessary impact on the median.

Glossary

Casualty: a person killed or injured during a terrorist event. Note that the definition of an individual being injured or wounded may vary from country to country and among media reports as well. Often, the number of casualties of terrorist attacks refer to the sum of the number of killed and wounded. The term death casualties refers to the number of deaths excluding the number of wounded.

Magnitude: the magnitude or impact of a terrorist event is often measured as the number of people killed and/or wounded during a terrorist attack. If both killed and wounded are included, a terrorist attack that kills 2 persons and wounds one person (3 casualties) would have a lower magnitude than a terrorist attack that kills 1 person and wounds 10 people (11 casualties).

Alternatively, one may consider that the number of fatalities alone better represents the magnitude of a terrorist attack. From this standpoint, a terrorist attack that kills 2 persons and wounds one person (2 death casualties) would have a higher magnitude than a terrorist attack that kills one person and wounds 10 people (1 death casualty).

ISIL: Islamic State of Iraq and the Levant (ISIL) is a Iraq-based terrorist group also known as the Islamic State, the Islamic State in Iraq, the Islamic State of Iraq and al Sham (ISIS), or Al-Qaeda in Iraq, which was founded in 2004. The main goal of ISIL is to create an Islamic caliphate across Iraq and Syria [4].

Taliban: ("students" in Pashto) is an Afghanistan-based terrorist group founded in 1994 during the Afghan civil war that followed the withdrawal of Soviet forces in Afghanistan in 1989. Its main goal is remove foreign military occupation in Afghanistan [4].

Boko Haram: is a Nigeria-based terrorist group founded in Nigeria in 2002, which became extremely violent in 2009. Its main goals are to remove traditional political and religious leaders and establish an Islamic government with strict Salafist interpretation of sharia in Nigeria [4].

Tehrik-i-Taliban Pakistan: (TTP) is a Pakistan-based terrorist group founded in Pakistan in 2007 by a shura council of 40 Taliban leaders. Its name translates to "Student movement" used in 1998 already, but unrelated to TTP. Its main goals are to apply sharia and combat the coalition forces in Afghanistan [4].

FURTHER READING

Stanton, J. (2013). Terrorism in the Context of Civil War. *The Journal of Politics*, 75(4):1009-1022

Frey, B. S. and Luechinger, S. (2005). *Measuring terrorism.* Law and the State: A Political Economy Approach: New Horizons in Law and Economics, Cheltenham, UK: Edward Elgar Publishing, 142-181.

LaFree, Gary and Dugan, Laura and Miller, Erin (2014). *Putting Terrorism in Context: Lessons from the Global Terrorism Database*. Abingdon, UK: Routledge.

CHAPTER 4

Myth No 3: The Vulnerability of the West to Terrorism

> Knowing that one may be
> subject to bias is one thing;
> being able to correct it is
> another.
>
> Jon Elster

4.1 Introduction: Asia and Africa in the line of fire

It is well established that individuals tend to make up their mind based on events that readily come to mind and underestimate or disregard those from a more distant memory. This form of cognitive bias, also known as "availability heuristic", is mainly influenced by media coverage [72].

Given the strong presence of global Western media in many countries across the world, events that occur in the West are more likely to be covered more extensively. Availability heuristics may therefore explain the discrepancy between popular belief and facts about the geographical coverage of terrorist events worldwide.

Evidence shows a different picture than what is often perceived in the West. Figure 4.1 shows that from 2002 to 2017 about 75% of the attacks occurred in Asia (including the Middle East), 15%

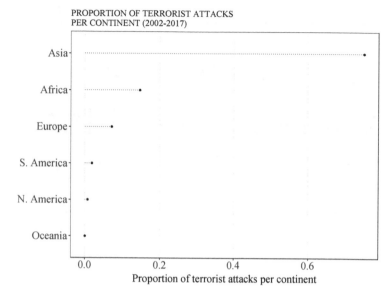

PROPORTION OF TERRORIST ATTACKS
PER CONTINENT (2002-2017)

Proportion of terrorist attacks per continent

Figure 4.1: Percentage of the attacks (*black dots*) per continent from 2002 to 2017. From *top* to *bottom*: Asia, Africa, Europe, South America, North America, and Oceania. Data source: GTD [56].

in Africa, 7% in Europe, 2% in South America, and about 1% in North America and Oceania together.

Not only were terrorist attacks from 2002 to 2017 mostly located in Asia and Africa, so were their victims. Terrorist attacks have provoked considerably more deaths in Asia and in Africa than anywhere else, as illustrated in Figure 4.2.

From 2002 to 2017, more than 140,000 individuals lost their life in Asia because of terrorist attacks. In Africa, terrorism killed more than 43,000 people, while in Europe, about 5,000 people died from terrorism. Over the same period, the total number of deaths in South America, North America and Oceania together was about 2,000 people.

4.2 One quarter of all attacks worldwide occur in Iraq

Close inspection shows that terrorist events are concentrated in specific countries in the Middle East and in few African countries. The top five countries with the highest number of attacks are: Iraq, Pakistan, India, Afghanistan, and the Philippines (Figure 4.3).

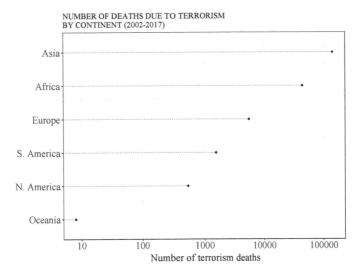

Figure 4.2: Number of deaths (*black dots*) caused by terrorist attacks per continent from 2002 to 2017 . From *top* to *bottom*: Asia, Africa, Europe, South America, North America, and Oceania. *Ticks placed on Log$_{10}$ scale.* Data source: GTD [56].

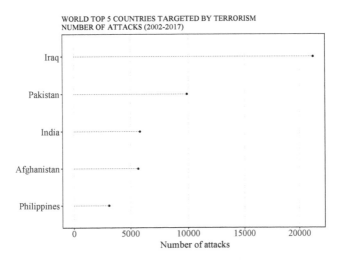

Figure 4.3: World top 5 countries with the highest number of terrorist attacks. Number of attacks (*black dots*) per country from 2002 to 2017. From *top* to *bottom*: Iraq, Pakistan, India, Afghanistan, and the Philippines. Data source: GTD [56].

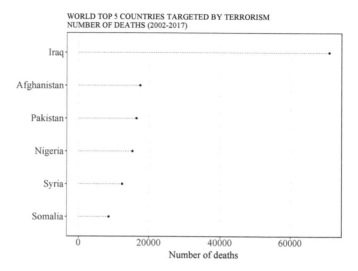

Figure 4.4: World top 5 countries with the highest death toll due to terrorism. Number of deaths (*black dots*) per country from 2002 to 2017. From *top* to *bottom*: Iraq, Afghanistan, Pakistan, Nigeria, Syria, and Somalia. Data source: GTD [56].

More than one quarter of attacks occurred in Iraq alone, which counts a total of 21,235 attacks from 2002 to 2017.

Of a total of 192,859 deaths due to terrorism worldwide from 2002 to 2017, Iraq counts more than 40% (79,596) of the total number of deaths worldwide. Figure 4.4 shows world's top five countries with the highest death toll due to terrorism.

Among the world top 20 countries with the largest number of lethal attacks (Table 4.1), half of them are in Asia, and a quarter of the countries are in Africa. There are only two countries in Europe (Russia and Ukraine), one in South America (Colombia), but none from North America or Oceania.

Among 124 countries that encountered lethal attacks from 2002 to 2017, the United States appears at the 36th rank with a total of 72 lethal attacks. The United Kingdom is placed at the 44th rank with a total of 33 lethal attacks, followed by France at the 45th position with 31 lethal attacks.

From 2002 to 2017, the United States encountered less than 0.5% of the lethal attacks that occurred in Iraq, less than 3.7% of the deadly attacks that occurred in Somalia, or about 6.2% of the number of lethal attacks experienced in Yemen.

Table 4.1: World top 20 countries with the highest number of lethal terrorist attacks observed from 2002 to 2017. Note that Russia and Turkey territories lie in Europe and Asia and count their largest number of attacks in Asia. Data source: GTD [56].

Rank	Country	Continent	Number of lethal attacks
1	Iraq	Asia	14563
2	Pakistan	Asia	4435
3	Afghanistan	Asia	3780
4	India	Asia	2031
5	Somalia	Africa	1992
6	Nigeria	Africa	1721
7	Philippines	Asia	1338
8	Yemen	Asia	1161
9	Syria	Asia	1118
10	Russia	Europe	589
11	Libya	Africa	580
12	Algeria	Africa	466
13	Colombia	S. America	448
14	Thailand	Asia	441
15	Turkey	Asia	431
16	Ukraine	Europe	369
17	West Bank and Gaza Strip	Asia	300
18	Sri Lanka	Asia	296
19	Democratic Republic of the Congo	Africa	264
20	Kenya	Africa	237

4.3 The most targeted city by terrorism: Baghdad, Iraq

At sub-national level, we observe that few cities have been extremely affected by lethal terrorism (attacks with at least one death). From 2002 to 2017, the top five most targeted cities by lethal terrorism are all located in the Middle East or in Africa.

The following most targeted cities include: Mosul (Iraq), Karachi (Pakistan), Mogadishu (Somalia), and Baqubah (Iraq) (Figure 4.5).

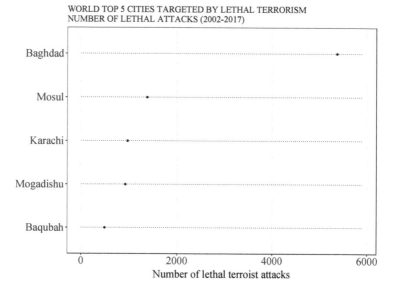

WORLD TOP 5 CITIES TARGETED BY LETHAL TERRORISM
NUMBER OF LETHAL ATTACKS (2002-2017)

Number of lethal terroist attacks

Figure 4.5: Total number of deadly attacks in the five most targeted cities in the world from 2002 to 2017. From *top* to *bottom*: Baghdad (Iraq), Mosul (Iraq), Karachi (Pakistan), Mogadishu (Somalia), and Baqubah (Iraq). Data source: GTD [56].

Baghdad, Iraq appears the most targeted city by lethal terrorism in the world. It encountered 5,372 deadly attacks, which corresponds roughly to 14% of all deadly attacks that occurred in the world (39,751) during the same period (2002-2017)! In other words, for every 7 deadly attacks that occurred in the world, one attack took place in Baghdad.

The observed types of terrorist attack and their relative frequency are peculiar to each city. Table 4.2 illustrates in further detail the number and relative frequency of attacks decomposed into nine types of terrorist attacks that occurred in Baghdad, Iraq and Paris, France from 2002 to 2017. It includes the following types of attacks: "Armed Assault", "Assassination", "Bombing or Explosion", "Facility or Infrastructure Attack", "Hijacking", "Hostage Taking (Barricade Incident)", "Hostage Taking (Kidnapping)", and "Unarmed Assault". If the type of attack is not known, it is classified as "Unknown".

While it is expected that Baghdad tends to encounter more attacks (in absolute terms) than Paris for most types of attacks—

Table 4.2: Comparing the type of terrorist attacks that occurred in Baghdad, Iraq and Paris, France from 2002 to 2017. Data source: GTD [56].

Type of attack	Number of lethal attacks	Frequency [%]	City
Armed Assault	719	10	Baghdad
Armed Assault	16	39	Paris
Assassination	354	5	Baghdad
Assassination	1	2	Paris
Bombing/Explosion	6201	84	Baghdad
Bombing/Explosion	10	24	Paris
Facility Infrastructure	14	0	Baghdad
Facility Infrastructure	6	15	Paris
Hijacking	0	0	Baghdad
Hijacking	0	0	Paris
Hostage Taking (Barricade Incident)	6	0	Baghdad
Hostage Taking (Barricade Incident)	4	10	Paris
Hostage Taking (Kidnapping)	97	1	Baghdad
Hostage Taking (Kidnapping)	0	0	Paris
Unarmed Assault	2	0	Baghdad
Unarmed Assault	4	10	Paris
Unknown	21	0	Baghdad
Unknown	0	0	Paris

this is always true except for the number of unarmed assaults—an analysis focused on the relative proportion of the types of attack can provide insight into the differences in the *modus operandi* of terrorist groups.

From 2002 to 2017, Baghdad was more prone to encounter bombing and explosions than Paris. More than 80% of attacks in Baghdad were bombing or explosions while this category of attacks represents less than one quarter of the attacks that occurred in Paris. Terrorist attacks in the French capital were more often carried out through armed assault—more than one third of the total number of attacks in Paris—compared to Baghdad, which encountered less than 10% of attacks of this type.

4.4   Conclusion: The most vulnerable regions to terrorism are in Asia and Africa

In this chapter, we have explored the spatial distribution of terrorist attacks and the associated number of deaths at continent level, and highlighted the countries and cities which are the most affected by terrorism.

Europe and North America are not more affected by terrorism than the other continents. On the contrary, the data show that Asia and Africa are by far more impacted by terrorism than any other continent, including Europe and North America, with regard to both the number of attacks and the associated deadly casualties.

At country-level, few countries appear by far more targeted by lethal terrorism than any other country. The five most affected countries (Iraq, Afghanistan, Pakistan, Nigeria, Syria, and Somalia) are all located in Asia or Africa. Among them, Iraq shows the highest number of deadly attacks with more than 3 times the number of lethal attacks observed in Afghanistan; the latter is ranked as the world's second most affected country by lethal terrorism.

Three of the top five most targeted cities by lethal terrorism are in Iraq. This includes Baghdad, the world's most targeted city by lethal terrorism, and the cities of Mosul and Baqubah. The Iraqi capital encountered 5,372 deadly attacks in 6 years, from 2002 and 2017.

In other words, the city of Baghdad alone encountered on average 335 lethal attacks a year, which is roughly a bit less than one lethal attack every day on average! The intensity of the lethal attack observed in this city is far above any intensity level of terrorism observed in North America or Europe.

The observations made from the exploration of GTD data brought light into a fundamental aspect of terrorism. Terrorism, and more particularly, lethal terrorism is mainly concentrated in few countries, and some cities are by far dominating the statistics. The following odd ratio summarizes the disproportion of terrorism that is observed in few areas compared to the world's average: *for every 7 terrorist attacks that kills someone in the world, one of them occurs in the capital of Iraq: Baghdad.*

The results of the analysis should be interpreted with a caveat. As mentioned in Chapter 2, we remind the reader that the empirical analysis carried out throughout this chapter and the following ones is based on data from one provider only. Therefore, their validity is constrained by the limitations of the data and a peculiar view on terrorism that follows the definition provided by the data provider. We cannot exclude that similar analysis carried out with another database would produce different results.

Media and terrorism: the double-edge sword

Globalization may engender two antagonistic effects on terrorism. Globalized terrorism may generate long-term destabilization, but conversely, may strengthen cooperation mechanisms among states, which are required to combat international terrorism [32].

Equally ambiguous, media coverage of terrorism is a "double-edged sword". Terrorists benefit from publicity and the attention that they desire, however it may also be damaging for them [63, p. 194].

"The relationship between publicity and terror is indeed paradoxical and complicated. Publicity focuses attention on a group, strengthening its morale and helping to attract recruits and sympathizers. But publicity is pernicious to the terrorist group too. It helps an outraged public to mobilize its vast resources and produces information that the public needs to pierce the veil of secrecy all terrorist groups require" [119].

Glossary

modus operandi: a concept derived from Latin that can be translated as mode of operating, and generally refers to a particular way of doing something [22]. Applied to terrorism activity, it may refer to the techniques, tactics, and/or procedures used by terrorist groups.

availability heuristic: a cognitive bias that may lead people to weigh more their opinion towards events that are easier to remember—often drawn from the most recent information they gather [72].

FURTHER READING

National Consortium for the Study of Terrorism and Responses to Terrorism (START) (2019). *see reports on the Global Terrorism Database.* https://www.start.umd.edu/gtd.

Rapoport, D. C. (1996). Editorial: The media and terrorism: Implications of the Unabomber case. *Terrorism and Political Violence,* 1(8): 7–9.

CHAPTER 5

Myth No 4: An Homogeneous Increase of Terrorism Over Time

> Absence of evidence is not evidence of absence.
>
> Martin John Rees

5.1 Introduction: identifying terrorism trends beyond visualization

This chapter aims to investigate if terrorism has increased over time, and if so, in which time period and geographical areas has the increase of terrorism been observed. We will explore temporal changes of terrorism with data from 2002 to 2017 and observe their evolution worldwide, and zoom into finer spatial level analysis to identify patterns at the level of continent, country, and city.

The work presented in this chapter will focus on two metrics of terrorism aggregated on a monthly basis: the number of events and its associated number of fatalities. Our analysis will apply statistical tools complementary to visual inspection of the data in order to identify and distinguish trends from other components that can explain the observed variation of terrorism in time.

Does a simple visualization of time series data suffice to make valid claims about trend in the data? In order to answer this question, we need to introduce a few concepts that relate to the analysis of data ordered in time, commonly referred to as time series data.

Times series data can be decomposed into three or more components, which can be multiplied or added. The *trend* component of a time series informs us on deterministic long-term behavior in the series; a positive trend term may reveal an upward pattern. The trend should be distinguished from other mechanisms that might causes variations in the time series without necessarily affecting the trend.

For example, in regions of continental climate, one may observe a progressive decrease in the temperatures from mid-summer until mid-winter, which coincides with a negative trend in the data. However, one may also observe short-term variations, such as up and down of the temperature between day and night. This regular pattern can be described through a *cyclical* component often associated with a sinusoidal (also called "wavelike") function. Cyclical monthly fluctuation often refers to seasonality.

Changes in temperature are also caused by phenomena that are virtually impossible to predict, such as micro variations of the wind direction and magnitude that operate at very fine spatial and temporal scales. The effects on temperature caused by random perturbations is described by a "noise" or "irregular" component.

All changes in the series that are not determined by the trend and/or cyclical components are usually captured by the noise component [41]. A typical objective in predicting time series data is to model all components, with the "noise" component that remains unexplained. In this framework, the noise values are assumed to be not correlated with each other and follow a Gaussian probability with zero mean and finite variance. In some research areas, this particular form of noise refers to "white noise".

A significant cyclical component could reveal the presence of high levels of terrorist activity followed by lull periods. The causes of observed cycles in terrorism may stem from a repetitive sequence of terrorist actions followed by government's responses.

Following a series of terrorist attacks, public opinion may push governments to take counterterrorism actions, which can temporarily reduce terrorism activity. In the meantime, terrorists may seize the opportunity to recruit new members and organize fresh attacks, which can lead to cycles of violence [1].

A better understanding of the cycles helps authorities to anticipate potential seasonal peaks of terrorism. This allows for better resource allocation: defensive measures can be increased during peaks and reduced during troughs [126].

In this chapter, we will explore and illustrate the observed number of monthly events and the associated number of fatalities along with the trend from 2002 to 2017, and this, at various spatial scales (continent, country, city). Throughout the text that follows, we will use *dashed* lines to indicate the observed data (aggregated at monthly level) and *thick plain* lines to highlight the trend component in the data.

Using the R package `forecast`, we extract the *trend* for a given month using a *two-sided* weighted moving average smoother based on values from each of the 6 previous months to each of the next 6 months, centered on the investigated month [112]. This procedure is also known as $2 \times MA$, for two moving averages [64][1].

5.2   Rise of terrorism in Asia and Africa

Globally, the monthly average number of terrorist attacks worldwide (*green thick line*, Figure 5.1) shows a positive trend from 2002 to 2015 while a reduction of the number of monthly events is observed from 2014 to 2017.

Despite the decrease from 2014 to 2017, worldwide terrorism activity remains at relatively high levels compared to the previous period (2002-2014). A similar pattern is observed in Asia (*gold thick line*) and similarly, to a lesser extent, in Africa (*black thick line* and in Europe (*blue thick line*).

The monthly number of terrorist attacks in Europe (*blue lines*) is consistently smaller over time compared to the monthly number of attacks in Asia (*gold lines*). The peak observed between 2013 and 2016 in Europe remains below the levels of terrorism observed in Asia and Africa. However, the trend in the evolution of the number of terrorist attacks in Europe shows an analogous pattern to what

---

[1]The approach can be summarized as follows: $x_t = \frac{1}{24}x_{t-6} + \frac{1}{12}x_{t-5} + \frac{1}{12}x_{t-4} + \cdots + \frac{1}{12}x_{t+4} + \frac{1}{12}x_{t+5} + \frac{1}{24}x_{t+6}$, with x the measured value of the phenomenon of interest (e.g. the number of attacks), which is computed for each month t. A small weight $\frac{1}{24}$ is applied to values at the extreme of the interval (for months $t-6$ and $t+6$) and a larger weight $\frac{1}{12}$ is applied to all other months $t-5$ to $t+5$ (except t) [64].

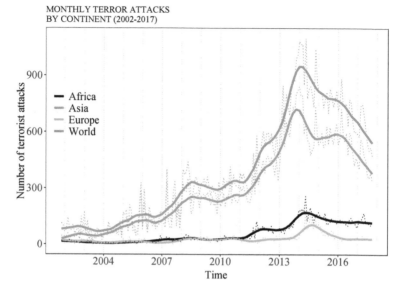

Figure 5.1: Monthly number of terrorist attacks (*dotted* line) and trend (*thick plain* line) perpetrated from 2002 to 2017 observed for the following continents: Asia (*gold*), Africa (*black*), Europe (*blue*), and the world (*green*). Data source: GTD [56].

is observed in Asia and Africa, with an increase of attacks focused between 2013 and 2016 followed by a relative decrease until 2017.

Recall from Chapter 4 that Asia counts the largest number of attacks among all continents. Its temporal pattern (Figure 5.1, *gold thick line*) closely follows worldwide aggregated values (*green thick line*). Furthermore, the Asian continent is consistently more targeted by terrorism than any other continents over time. The results are not surprising since Asia includes the most targeted countries in the world such as Iraq, Afghanistan, Pakistan, and India.

Africa also shows relatively high number of monthly attacks. This continent counts various countries that have been highly targeted by terrorism, such as Nigeria. Since 2009, the northeast of Nigeria has been devastated by terrorist attacks perpetrated by Boko Haram, which has led to thousands of deaths and several millions of refugees that fled their homeland [140]. Nevertheless, Africa consistently shows a lower number of monthly attacks than Asia over time.

5.3   No temporal pattern in the West?

Figure 5.2 shows the evolution of the monthly number of terrorist attacks in North America (*black lines*), South America (*blue lines*), and Oceania (*gold lines*). We show the results for these continents separately from Asia, Africa, and Europe (Figure 5.1) for the sake of clarity. The monthly number of attacks in the latter continents is much smaller than the former, as illustrated by different orders of magnitude of the y-axis in Figure 5.1 compared to those in Figure 5.2.

In contrast, the monthly number of attacks (Figure 5.2) in South (*blue thick line*) and North (*black thick line*) America, and in Oceania (*gold thick line*) does not suggest any particular temporal tendency. These regions have encountered relatively low levels of terrorism compared to the other continents and therefore exhibit relatively small variations in the data.

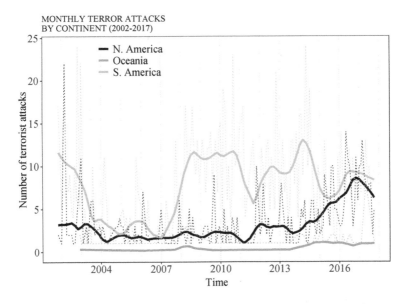

Figure 5.2: Monthly number of terrorist attacks (*dotted* line) and trend (*thick plain* line) perpetrated from 2002 to 2017 observed for the following continents: North America (*black*), Oceania (*gold*), and South America (*blue*), and the world (*green*). Data source: GTD [56].

5.4   Rise of deadly casualties in Asia and Africa

Consistently, the monthly average number of deaths generated by terrorism (Figure 5.3) suggests a positive trend in the world (*green thick line*), Asia (*gold thick line*) and Africa (*black thick line*), mainly between 2010 and 2014, followed by a relative decrease until 2017.

Despite the observed decrease from 2014 to 2017, the number of deadly casualties in the last two years remains at high levels compared to the values observed between 2002 and 2014. The tendency in Europe (*blue thick line*) is less marked, although it shows relative high intensity and variability in the number of attacks observed between 2013 and 2016.

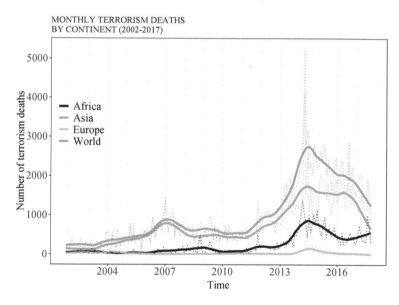

Figure 5.3: Monthly number of terrorism deaths (*dotted* line) and trend (*thick plain* line) associated with attacks perpetrated from 2002 to 2017 observed for the following continents: Asia (*gold*), Africa (*black*), Europe (*blue*), and the world (*green*). Data source: GTD [56].

5.5   No temporal pattern in terrorism deaths in the Americas and Oceania?

Figure 5.4 shows the evolution of the monthly number of deadly casualties in South America (*blue lines*), North America (*black lines*), and Oceania (*gold lines*).

In contrast to Asia, Africa, and Europe, there is no evidence of a temporal pattern in the number of deaths due to terrorism in South America (*blue thick line*), North America (*black thick line*), and Oceania (*gold thick line*) from 2002 to 2017 (Figure 5.4).

5.6   High levels of terrorism persist in very few countries

The direction and magnitude of the temporal trend of time series data may vary according to the spatial level in which data is aggregated. For example, a positive trend observed between 2010 and 2014 in Asia does not mean that Japan exhibits a similar trend.

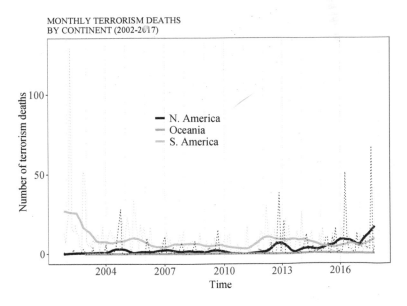

Figure 5.4: Monthly number of terrorism deaths (*dotted* line) and trend (*thick plain* line) associated with attacks perpetrated from 2002 to 2017 observed for the following continents: Oceania (*gold*), North America (*black*), and South America (*blue*). Data source: GTD [56].

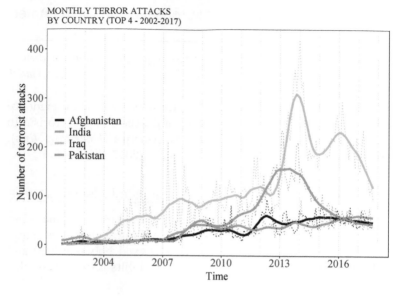

Figure 5.5: Monthly number of terrorist attacks (*dotted* line) and trend (*thick plain* line) in the top 4 most targeted countries in the world from 2002 to 2017: Iraq (*blue*), Pakistan (*green*), India (*gold*), and Afghanistan (*black*). Data source: GTD [56].

The analysis of data aggregated at finer scales may help refine our understanding about the processes at stake. In this section, we explore the temporal pattern of the monthly number of terrorist attacks and associated deadly casualties at country-level. We focus on the world's four most affected countries by terrorist events from 2002 to 2017: Iraq, Pakistan, India, and Afghanistan.

The data illustrated in Figure 5.5, Iraq (*blue thick line*) and Pakistan (*green thick line*) suggest an increasing trend of the monthly number of terrorist attacks until about 2014, followed by a relative decrease in the number of attacks in 2017. The number of attacks observed in Iraq remains at higher levels from 2015 to 2017 compared to most months in the previous time period (2002 to 2014).

However, the number of attacks observed in Pakistan (*green thick line*) has reduced drastically after 2014. In Afghanistan (*black thick line*) and India (*gold thick line*), a slight tendency of an increase in terrorism is observed between 2002 and 2017. However, one cannot exclude the absence of temporal pattern given that the high relative variability in the data.

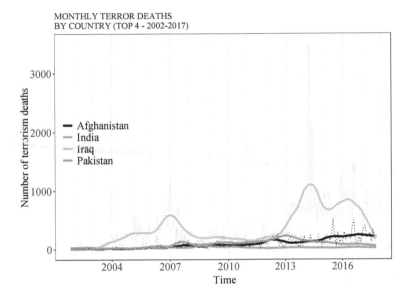

Figure 5.6: Monthly number of terrorism deaths (*dotted* line) and trend (*thick plain* line) in the top 4 most targeted countries in the world from 2002 to 2017: Iraq (*blue*), Pakistan (*green*), India (*gold*), and Afghanistan (*black*). Data source: GTD [56].

The number of deaths due to terrorist attacks is highlighted in Figure 5.6 for Pakistan (*green lines*), Afghanistan (*black lines*), India (*gold lines*), and Iraq (*blue lines*). In Iraq, peaks of deadly casualties are mainly observed in the later years (2014, 2016, 2017), while high variation is observed during the investigated time period. More deaths have been generated in Iraq in the more recent time period (2013-2017) compared to 2002-2012, although local peaks can be observed in 2007 for example.

In Pakistan (*green thick line*), Afghanistan (*black thick line*), and India (*gold thick line*), the temporal pattern is less clear. One cannot exclude the absence of temporal pattern in the monthly number of deaths due to terrorism in these countries given that the values do not vary drastically over the investigated time period.

5.7 Dynamics of terror events and death toll in the world's most targeted city

Changes in the number of events and associated deaths are the results of various factors that include the resources, strategies, and tactics employed by terrorist groups as well as their interaction with counterterrorism activity that may also vary over time and space.

The set of factors involved in terrorism and their effects may also vary in relative fine spatial scale. Terrorist groups tend to target very specific locations. Terrorist attacks are not homogeneous in space. In other words, within a given country, the intensity of terrorism can drastically vary from one city to another. In Chapter 6, we describe in further details how terrorism targeting results in clusters that can be observed and analyzed.

In this section, we focus our exploratory analysis on the world's most targeted city by terrorism: Baghdad, Iraq. The Iraqi capital has been regularly affected by high intensity activity of terrorism, especially during the Iraq war (2003-2011), which started with the U.S.-led military invasion of Iraq 19th March 2013. The city of Baghdad became under control of the US coalition forces 10th April 2013.

Following the Iraq war, the country encountered a period of growing insurgency violence which led to civil war that started in January 2014. During the first years of the civil war, Baghdad was strongly affected by terrorist attacks perpetrated by the Islamic State of Iraq and the Levant (ISIL). The terrorist group took control of the city of Mosul and Tikrit in early June 2014. The end of the civil war in December 2017 coincides with the fall of ISIL. The group lost more 95% of its territory [7, 149].

Figures 5.7 and 5.8 adds the context and important events in the city of Baghdad associated with changes in terrorism activity and associated deaths, respectively observed from 2003 to 2017. In Figure 5.7, we observe high variations over the months and notice an increase in the monthly number of attacks from 2013 to 2016. As illustrated in Figure 5.8, there are high variations in terrorism activity in Baghdad. After the Fall of Baghdad, the number of deaths increase until 2017, with peaks between 2006 and 2007. Between the Iraq War and the Iraq Civil War, the number of terrorism deaths increase and exhibit several local peaks between 2013 and 2016.

Baghdad exhibits the largest number of terrorist attacks and associated deaths worldwide. The city was at war about 85% of the time during our investigated time period, from 2003 to 2017. The strategic position of Baghdad during a war context is one among numerous factors that explain the extremely hight levels of terrorism intensity observed in this city.

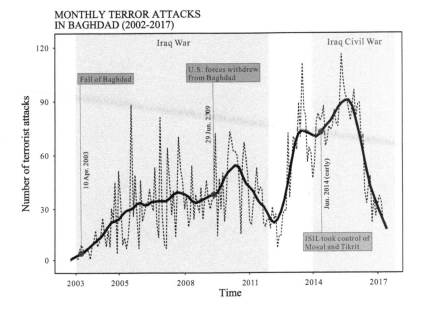

Figure 5.7: Monthly number of terrorist attacks (*dotted* line) and trend (*thick plain* line) in Baghdad, Iraq from 2003 to 2017. The Iraq war (2003-2011) and Iraq civil war (2014-2017) are represented by shaded areas. Data source: GTD [56].

5.8   Conclusion: an uneven temporal variability of terrorism across continents, countries, and cities

In this chapter, we have first explored temporal evolution of the number of attacks and its associated number of deaths using monthly GTD data from 2002 and 2017. We analyzed data at various spatial scales in order to detect trends that may occur worldwide and at continent level. We further investigated temporal changes of terrorism in the world's most targeted countries and city.

It would be inaccurate to claim that terrorism has increased everywhere from 2002 to 2017. General statement about the temporal pattern of terrorism require careful attention. First, temporal trends vary according to temporal level of aggregation. In our study, we aggregate data at a monthly level. Other temporal levels of aggregation could lead to different results.

MONTHLY TERROR DEATHS
IN BAGHDAD (2002-2017)

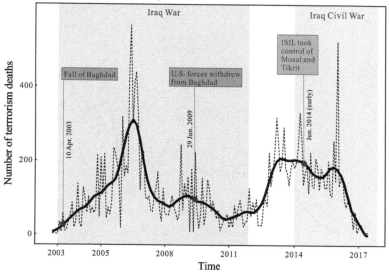

Figure 5.8: Monthly number of terrorism deaths (*dotted* line) and trend (*thick plain* line) in Baghdad, Iraq from 2003 to 2017. The Iraq war (2003-2011) and Iraq civil war (2014-2017) are represented by shaded areas. Data source: GTD [56].

Second, caution is required when comparing 2011 and 2012 GTD events. The number of attacks and fatalities are underestimated before 2012. Further details on the methodological changes performed by GTD in 2012 and their impact on the interpretation of results are discussed here: https://www.start.umd.edu/news/ discussion-point-benefits- and-drawbacks-methodological -advancements-data- collection-and-coding.

Third, both the region of investigation and the spatial level of analysis affect the observed temporal trend. Temporal patterns may differ if one considers the worldwide average or if one focuses on data aggregated at continental, country, or subnational level. For example, a positive trend of the number of attacks observed in Baghdad between 2014 and 2016 may not coincide with the observed pattern in Iraq or in Asia.

Fourth, an increase in the number of terrorist attacks may not be necessarily associated with an increase in the number of deadly casualties for a given time period. For example, despite being rare, mass-casualty attacks may drastically increase the number of deaths observed in a given month while the increase in the count of terrorist events would be equal to an attack that would have generated few or no deadly casualties.

While general statements about the temporal pattern of terrorism remain problematic, one can observe a few important trends that are consistent over time and observed at various spatial scales. From 2002 to 2015, both the monthly number attacks and the associated monthly number of deaths seemed to have increased globally.

We observed a positive trend from 2010 to 2014 particularly marked in the world and in Asia. The temporal pattern of the number of attacks and deaths due to terrorism in Asia closely follows worldwide trends, since most attacks occurred in Asia, which includes Iraq, Pakistan, and Afghanistan. From 2015 to 2017 a negative trend follows in both the monthly number of attacks and death toll in the world and mainly in Asia.

Evidence from the data did not allow us to identify any particular temporal pattern with regard to the monthly number of attacks and associated deaths in South America, North America, and Oceania during the investigated time period. However, we observed that over time both the monthly number of attacks and the associated number of deaths have consistently exhibited lower levels compared to those observed in Asia, Africa, and to a lesser extent the levels observed in Europe.

From 2002 to 2017, the data show that over time Asia and Africa have been consistently more impacted by terrorism than Europe, North America, South America and Oceania with regard to both the number of attacks and the associated deadly casualties. This result provides support to the findings described in Chapter 4, which suggest that on average (2002-2017), Asia and Africa remain the most targeted continents by terrorism.

At country-level, a few countries are disproportionately targeted by terrorism. This includes Iraq, Pakistan, India, and Afghanistan. Our analysis suggested a positive trend in the number of attacks and associated deaths in Iraq until 2015 followed by a relative decrease after 2015. However, the intensity of attacks (number of attacks and deaths) remains high. The number of events and deaths in Pakistan, India, and Afghanistan is smaller than those observed in Iraq. Also, their temporal trends are less marked compared to what we observed in Iraq.

In addition, we identified interesting features in the monthly number of attacks and deaths from 2002 to 2017 in Baghdad, Iraq.

The Iraqi capital was the world's most targeted city by terrorist events during this period, and also more specifically, by lethal terrorism. We showed that peaks in the number of deaths may not necessarily coincide with larger number of events. A mass-casualty attack may lead to a drastic increase in the monthly number of deaths while it counts as one event.

The analysis made in this present chapter informs us that temporal trends can vary according to the spatial and temporal domains and scales on which terrorism is observed. Furthermore, the metrics (number of attacks, number of deaths) used to analyze the magnitude of terrorism reveal different facets of terrorism.

Consequently, the validity of inference made on the analysis of temporal patterns of terrorism remains limited. Reliable research work should acknowledge the source and limitations of: (i) the data (e.g. GTD, or other source), (ii): terrorism metric (e.g. number of attacks, number of deaths), (iii) temporal (e.g. monthly, weekly data), and spatial level of aggregation (e.g. city, province, country), and (iv) the temporal (e.g. several years, one month) and spatial domains (e.g. continent, one country).

Thus, our results should be interpreted with a caveat. We sometimes did not find evidence of patterns; however, this cannot be interpreted as evidence of the absence of patterns. This can be described in the words of Jon Elster quoted in the beginning of this chapter: "absence of evidence is not evidence of absence". Furthermore, we cannot exclude that the same analysis using different data, terrorism metrics, and levels of aggregation or different spatio-temporal domains would produce similar results.

Time series analysis: the science of modelling dynamics

Time series analysis could be defined as the science of modeling dynamics: things that change over time. In this framework, the phenomenon under investigation is represented by a time series, defined as a sequence of values of a random variable typically taken at successive equally spaced discrete intervals over time [60, p. 1].

In the Western world, early work on time series analysis go back to the 17th century with the analysis of games of chance by Blaise Pascal and Pierre de Fermat, and profit from analysis by John Graunt and other merchants [75]. For a thorough review of the most common time series models still used nowadays, see e.g. [60].

Glossary

Moving average: a (simple) moving average is the (arithmetic) succession of averages computed on successive time periods (usually of same size and with overlapping between the periods) of a time series.

Noise component: also called "irregular" component which describes random variations of the data as deviations of the observed time series from the underlying pattern.

Seasonal component: wavelike movements, quasi regular fluctuations around the long-term trend, lasting for long periods.

Time series components: a time series is often composed of a trend, seasonal (and/or cyclical), and noise components, which can be combined in different ways (e.g. in an additive or multiplicative way). Accordingly, to describe the observed data.

Trend component: long-term, gradual increase or decrease of the random variable associated with the observed phenomenon.

FURTHER READING

Harvey, Andrew C (1993). Time Series Models. *MIT Press,* Second Edition. Cambridge, MA, USA.

CHAPTER 6

Myth No 5:
Terrorism Occurs
Randomly

> I group random with
> stochastic or chancy, taking
> a random process to be one
> which does not operate
> wholly capriciously or
> haphazardly but in accord
> with stochastic or
> probabilistic laws.
>
> John Earman

6.1 Introduction: spatial patterns of terrorism rely on spatial scales and lenses to view spatial data

Terrorist attacks may indiscriminately target civilians and often appear as if they target locations and people at random. Consequently, terrorist acts are often depicted by the media and perceived by citizens as "random" events. However, do terrorist events really occur by chance? How can statistics help answer this question?

In this chapter, we will focus on the location of terrorism in space and provide further clarification on the statistical concept of spatial randomness. We will discuss how terrorism can be viewed as a spatial phenomenon within a statistical framework and introduce statistical tools to analyze its properties. We will show that contrary to popular belief, terrorist events often occur close to each other, and hence, form clusters at various spatial scales.

The detection of spatial pattern of terrorism depends on various assumptions. We will show throughout the chapter how the theoretical lenses by which spatial data is assumed to be generated may affect the results of the analysis of spatial patterns. Furthermore, we will illustrate the important role of the geographic scale upon which the detection of clusters depends.

We will conclude the chapter by pointing out important limits of the spatial analysis of terrorism data. We will learn that the validity of the spatial analysis of global terrorism data is limited by the spatial accuracy of the data. A case study focusing on terrorist attacks that occurred in Baghdad from 2002 to 2017 will be introduced to illustrate important insights that can be obtained from the spatial statistical analysis of terrorism data while identifying the major limitations that affect the validity of the results and their interpretation.

6.2   Is terrorism spatially random?

It is undeniable that some terrorist attacks are indiscriminate on purpose. In specific contexts, some terrorist groups do not aim at targeting one or several specific individuals; rather, they deliberately aim at harming anybody present in the location of the attack.

The spatial locations of terrorist events—often represented by spatial coordinates such as the longitude and the latitude—are however never the result of pure chance [9]. Scholars are unanimous: terrorists do not commit actions randomly [63, 81]. They do not strike anywhere at any time [96].

Since terrorist attacks are not the results of pure chance, spatial patterns of terrorism can be detected at various scales using relevant statistical techniques. Hence, one can highlight areas that are more affected by terrorism than others within countries or within lower-level administrative regions. However, can we completely exclude spatial randomness of terrorist events at *all* spatial scales?

At very fine scale—for example if one considers processes that may affect the final location of terrorist events at meter or centimeter scale—randomness may prevail. Spatial patterns of terrorism may be very different from what is observed at macro scale (e.g.

among cities or regions in a country). Consider a case where a
bomb is supposed to detonate in a specific location, say at the
intersection of two roads, as carefully planned by a terrorist group.
Despite all efforts made by the group to dispose the bomb in a
precisely determined location, its exact coordinates, at centimeter
level, will depend on unobservable factors that cannot be determined in advance.

In principle, global terrorism is not investigated at spatial scales
below the perimeter of cities. This comes from the fact that the
database providers of global terrorism, such as GTD [56], do not
indicate the exact location of the attacks; rather, they provide the
spatial coordinates associated with the centroid of the city in which
the events occur (see Section 2.3).

Furthermore, the extent of spatial randomness present in the
data may depend on the targets of terrorist attacks. For example,
the location of attacks on moving targets, such as military per-
sonnel or public figures may, on average, exhibit spatial randomness across larger distances compared to attacks targeting fixed
infrastructures such as bombs placed on bridges or governmental
buildings.

Most literature on terrorism has analyzed terrorism at relatively
large scales, in order to compare terrorism activity between countries or among subnational regions within countries. If terrorism
data is aggregated at country-level, spatial randomness that may
operate within country can in principle be considered negligible
and can be ignored without further consequences on the results,
assuming that inference is not made below country-level.

However, it remains challenging to analyze the spatial patterns
of terrorism at finer scales, such as city-level. Terrorist events from
global terrorism databases exhibit a relative large uncertainty with
regard to their locations, which prevents statistical analysis from
capturing terrorist patterns at fine scales. In Section 6.7, we illustrate and provide further details on the obstacles to detecting spatial patterns of terrorist attacks within city boundaries.

6.3   Why should we care about spatial autocorrelation?

Terrorist events can be elegantly described through statistical models that depict them as spatial points. In this simplification of
the reality, terrorist events are summarized by their location (spatial coordinates) and additional features can be attributed to each
event, such as the number of casualties.

In the spatial statistics terminology, if the locations of the points (also called events) in a given investigated area are exclusively due to chance, they can be viewed as one realization of a *complete spatial random* (CSR) process. In this case, we consider the CSR as the underlying process that has generated these points.

There are numerous examples in nature where events appear "haphazardly" distributed in one, two, or more dimensions. This can be observed in time (one dimension) for the emission of radioactive particles in time, on a two-dimensional surface for the location of trees in a forest, or in a three-dimension space for the spatial distribution of stars observed in a portion of the sky.

Yet social phenomena that result from a CSR process are less frequent. Locations at close proximity tend to share similar characteristics that may affect social events. In addition, the presence of social events in a given location often attracts similar events in neighboring locations.

The correlation of a phenomenon with itself is called autocorrelation. Temporal autocorrelation refers to correlation over time, while spatial autocorrelation refers to correlation across space. The underlying concept behind spatial autocorrelation can be understood through Tobler's first law of geography: "everything is similar, but near things tend to be more similar" [141].

In the presence of spatial autocorrelation, the main assumptions of standard statistical models are violated and their results may not hold. More specifically, the difference between the observed values associated with the investigated variable and the predicted values—this difference is often called residuals—may not be assumed independent and identically distributed (i.i.d). Hence, errors of the Ordinary Least Square (OLS) regression—the OLS method is a widely used regression method which aims at minimizing the sum of square differences between the observed and predicted values— are spatially correlated and the estimated parameters (regression coefficients) are inefficient.

Spatial autocorrelation affects the interpretation of the results of a wide range of statistical models in addition to the OLS regression. The i.i.d assumption that ensures the validity of statistical tests which assume equal variability of the variance over the range of the predictors (homoskedasticity) are violated. Spatial autocorrelation may generate bias in standard Students' t-tests and measures of fit (Pearson's r) and lead to wrong inference [3].

As a remedy, scholars have developed methods to remove or mitigate the nuisance generated by spatial autocorrelation. By accounting for the spatial dependency present in the data, some methods applied to regression models improve the accuracy of the parameters estimation. Alternatively, one can study the spatial structure that results from spatial dependencies. Research that focuses on the spatial dependency in the data has allowed researchers in various applied fields to reveal the mechanisms behind the observed spatial patterns.

In a spatial statistical model, the spatial structure present in the data reflects the effect of unobserved factors on the variation of the dependent variable. Since it is virtually impossible to build a model that includes all factors that affect a social phenomenon, the information provided by the spatial structure can bring insight into unknown or/and unaccounted mechanisms. In the following section, we will discuss in further details the main approaches used to view and model spatial data and analyze its spatial dependency structure.

6.4   Choosing relevant lenses to explore spatial data

Spatial data can be viewed according to different "lenses", each of them, offering a specific view of the reality in which spatial phenomena are observed. These models are commonly classified into three groups: *point processes* (Figure 6.1), *lattice* (Figure 6.2) and *geostatistical* (Figure 6.3).

Choosing suitable lenses to view and model spatial data is not straightforward. The choice depends on the type of data, the model that describes the process that generated the data, the level of spatial aggregation, and the research question.

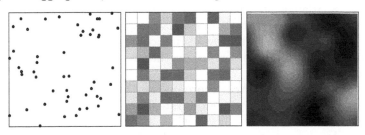

Figure 6.1: Realization of a point process

Figure 6.2: Realization of a lattice process

Figure 6.3: Realization of a geostatistical process

Lattice approaches are used to model spatially continuous or discrete data. Spatially continuous data are associated with phenomena that can occur anywhere in a given spatial domain of investigation. For example, temperature can be viewed as spatially continuous. It is reasonable (at least from a classical mechanics perspective) to assume that temperature can be measured everywhere. Spatially discrete data are typically associated with phenomena that are present in locations and absent in others. For example, trees occupy distinct areas in a forest. Characteristics of the trees such as height, or width cannot be measured everywhere in the forest but only at the location of the trees.

In lattice models, locations where the phenomenon is measured (reference locations) are defined on discrete spatial features such as pixels from a grid feature or irregular polygons such as administrative regions for example [31, p. 167]. A digital picture taken from your smart phone is an example of lattice data, in which each pixel is characterized by a value, which represents a specific color. All together, the pixels constitute the picture.

Lattice models can also accommodate discretized spatial domains composed of irregular polygons. Counties in the United Kingdom are example of irregular polygons. A lattice approach may be suitable to make statistical inference on terrorism based on the total number of attacks counted per county in the United Kingdom. In this framework, terrorist events are aggregated at county-level, therefore we assume that we only know the number of events per county and inference will ignore variations within county.

One may view terrorist events as dots drawn on a blank paper. Dots refer to discrete data called spatial points and the area on which they can be drawn is the spatial domain. In this framework, terrorist events are not aggregated into regular or irregular polygons. Spatial points can be characterized by the location of the corresponding terrorist attacks and the spatial domain can designate a country delimited by its national boundary for example. Hence, one can make inference on the processes that have generated the points.

Spatial point processes represent an elegant statistical framework to describe this type of discrete data. Within a spatial point process model approach, the locations of terrorist events is assumed to be the result of an underlying stochastic process. Hence, one may model and analyze the expected number of terrorist events per unit of area. The rate of terrorist events (count per spatial unit) refers to the intensity of the associated point process. In Section 7.5, we use a point process approach to analyze the intensity of terrorist events perpetrated by the Islamic State in Iraq.

Finally, geostatistical models cover random processes continuous in space [31, p. 124]. These models use geolocated data of a phenomenon of interest with values provided at fixed locations to predict values at unknown spatial locations. For example, one may assume that the propensity of terrorism to generate deaths (lethality) is continuous across space akin to that of temperature. The data are gathered from the location of terrorist events, which are considered fixed and used to reconstruct the unobserved (latent) process underlying the lethality of terrorism [115].

6.5 Terrorism is spatially correlated at various spatial scales

Evidence suggests that terrorist events are spatially correlated at various spatial scales [9]. One observes that areas closer to locations that have been targeted by terrorism are often more at risk than areas further away [115].

Spatial dependency is observed from local (e.g. across cities within a province) to global scales (e.g. across neighboring countries). One can examine a phenomenon at several spatial scales, and may identify different types of processes at stake for each investigated scale. This means that the results from a statistical analysis of spatial data are scale-dependent.

For example, if one studies the spatial distribution of terrorist attacks among provinces in Iraq, one may observe that there are areas that exhibit greater than average number of events in Baghdad province. These areas refer to spatial clusters [2], and in this case, are observed at province-level in Iraq.

Alternatively, one may examine the same phenomenon with data restricted to attacks that occurred in a popular street in Baghdad that is regularly under attack. Despite careful planning, it is unlikely that terrorist groups can determine in advance the exact location, say within a radius of few meters, of a bomb that is programmed to detonate in a particular area. For attacks on moving targets, for example attacks on law enforcement officers, the exact location of the attacks may be even more difficult to plan in advance.

Therefore, if one would analyze a large number of attacks in a relatively small area highly targeted by terrorist attacks, say in a given street, it is reasonable to expect that randomness may prevail. Clusters of terrorism that can be observed at coarser scales (regional, national level of analysis) can vanish at finer spatial scales. If so, one cannot distinguish the location of the attacks from those that would be the results of a random (CSR process).

The identification of terrorism clusters started with pioneered work in the 1980s that focused on how previous terrorist attacks influence future ones aiming at detecting factors that lead to high concentration of attacks in specific time periods or terrorism clusters in time [59, 54]. Later, researchers analyzed the spatial dimension of terrorism. They mainly focused on analyzing terrorism at country-level [11]. However, studies at that spatial scale ignored the inherent local nature of terrorism: terrorism tends to target specific locations; some locations are more at risk than others. There is no country affected by terrorism which would exhibit an equal risk of terrorism within its national border.

Consequently, recent research work has taken benefit from disaggregated data on terrorism to investigate terrorism at finer spatial levels. However, subnational spatial patterns of terrorism have been mainly investigated through descriptive or confirmatory analysis [109, 104, 83, 82, 52] which did not allow for understanding the processes behind the observed patterns. More recently, more advanced approaches that account for uncertainty have brought key insight into the mechanisms behind the clusters of terrorism observed at relatively fine temporal and spatial scales [94, 101, 115].

Point processes have been successfully used to model and understand the mechanisms that drive social phenomena in both space and time, such as crime [123, 85], conflict [85] and insurgency [152]. Point process approaches are particularly useful since they offer a rigorous framework to capture the scale(s)at which clusters of terrorism are observed. This can be carried out through the analysis of the spatial distribution of the data in comparison with those from a complete random process. The detection of point process clusters is illustrated in further details in the next section.

6.6   Spatial inaccuracy: what does that mean in practice?

Between 2002 and 2017, numerous terrorist attacks took place in various locations within urban areas of two Iraqi cities: Baghdad and Mosul. Over this time period, these cities were the world's most targeted cities by terrorism [56]. In order to illustrate the application of a point process to terrorism data, the following sections of this chapter will focus on the analysis of the spatial patterns of terrorism that occured in Mosul and Baghdad from 2002 to 2017.

Global terrorism data has shortcoming with regard to its spatial accuracy. GTD provides the geo-location of terrorist attacks corresponding to the centroid of the city in which events occur. This means that all terrorist attacks that take place in a city are labeled with the same coordinates. The pair of coordinates longitude-latitude provided by GTD is therefore identical for all attacks that occurred in the same city.

In Baghdad, this means that all 8,632 attacks that occurred from 2002 to 2017 are artificially concentrated into one spatial location: the centroid of Baghdad. Therefore, within-city heterogeneity of terrorist attacks cannot be observed and quantified if one only uses the geographical coordinates of the attacks provided by GTD.

In order to emulate within-city spatial variation of terrorism, we simulated the location of the attacks in urban areas (Figure 6.4: *gray polygons*) using urban extent polygons from NASA Socioeconomic Data and Applications Center (SEDAC) [14]. We used the R package sp, to simulate the point patterns generated by a CSR process with number of points in the urban areas corresponding to the number of observed events (*blue shades* areas in Figure 6.4), this for each city provided by GTD.

Figure 6.4 shows the number of terrorist attacks on a log scale (*blue shades*) that occurred from 2002 to 2017 aggregated within a 0.05 degree resolution grid. The urban areas and central points (centroid of the urban areas) of Baghdad and Mosul are highlighted with *red polygons* and *red points*, respectively.

Is it reasonable to assume that the locations of terrorist events follow a CSR process within municipality boundaries? How would spatial patterns of terrorist events from a CSR process look like? In the next section, we investigate this in further details and illustrate the spatial distribution of terrorist events within cities based on different point process approaches.

6.7 In the bull's eye!

The true locations of terrorist attacks within the city of Baghdad are not known. We simulate point patterns based on three different types of point processes in order to emulate several hypothetical spatial distributions of terrorist attack in Baghdad.

Figure 6.5 shows the centroid of Baghdad (*red disk*), as provided by GTD and the realization of simulated locations of terrorist attacks (*black crosses*) based on three point processes in the urban

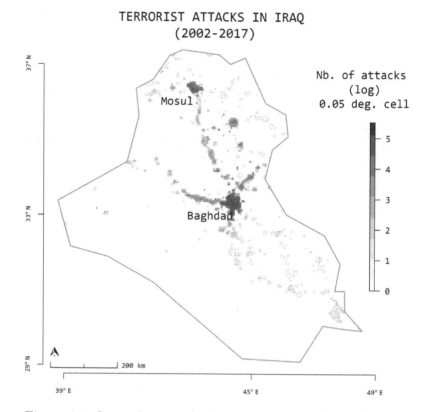

Figure 6.4: Count (log scale) of terrorist attacks (21,238) perpetrated from 2002 to 2017 (*blue gradient*) aggregated within 0.05 degree grid-cells in Iraq. Urban areas are delimited by *gray polygons*. The main targeted cities are Baghdad and Mosul (urban area: *red polygons*, centroid: *red points*) and their neighborhood. Data source: GTD [56]; urban polygon [14].

area of Baghdad (*red polygons*). *Left*: a complete spatial random (CSR) process; *center*: a clustered process; and *right*: a regular point process.

The realization of each point process is based on a total of 8,632 events, which correspond to the observed number of terrorist attacks that occurred in Baghdad from 2002 to 2017. As discussed in Section 6.2, a CSR process applied to terrorist events assumes that their locations are independent. This means that attacks do not influence each other and their locations are only the result of chance.

A CSR process can be modeled by the so-called spatial *homogeneous* Poisson process [74, pp. 1-3]. In this framework, the expected count of terrorist attacks per spatial unit is equal everywhere within the domain of investigation. Applied within the urban boundary of Baghdad, the process assumes that the risk of getting an attack is the same anywhere in the city.

However, it has been observed that terrorism tends to repeatedly target specific locations, and hence, aggregates in space and therefore exhibit positive spatial autocorrelation. It is therefore unlikely that the true location of terrorist attacks in Baghdad would be the result of a CSR process, as illustrated in Figure 6.5 (*left*).

Equally, observing a pattern similar to the regular pattern illustrated in Figure 6.5 (*right*) appears unlikely. This regular pattern represents one realization of a point process in which events tend to repulse each other. In this framework, their spatial autocorrelation is negative.

As a result, a regular process assumes that terrorist events are located as far as possible from each other. The result is that the spatial distribution of the terrorist attacks form regular patterns. This type of pattern is also unlikely to be observed in our case study given the clustering nature of terrorism observed in other contexts and at various spatial scales.

A clustered process, as illustrated in Figure 6.5 (*center*) is therefore more likely to be observed. It is reasonable to expect that some locations, e.g., US embassy, or military bases are more likely be highly targeted compared to others. The locations of the points have been obtained by simulation, however. This means that some important questions will remain unanswered.

Within the urban area of Baghdad, where are the most targeted areas? How many clusters are expected? How far do terrorist events influence each other? Which factors drive the clustering process?

This illustrative example aimed at pointing out that spatial analysis of terrorism based on GTD alone does not suffice to make within-city inference due to spatial inaccuracy of the reported events. Higher accuracy in the data is required to investigate patterns of terrorism inside municipality boundaries.

It is reasonable to expect within-city clusters in various cities targeted by terrorism worldwide but they cannot be captured with GTD data alone. However, spatial clusters of terrorism can be

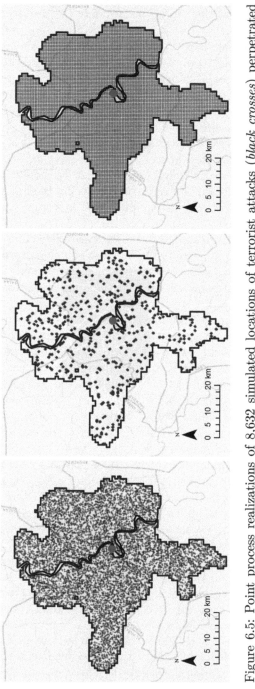

Figure 6.5: Point process realizations of 8,632 simulated locations of terrorist attacks (*black crosses*) perpetrated from 2002 to 2017 within the urban area of Baghdad, Iraq. *Left*: complete spatial random (CSR); *center*: clustered; and *right*: regular point process. The centroid of Baghdad provided by GTD is represented by a *red disk*. The urban area of Baghdad is delimited by a *black polygon*. Data source: GTD [56]; urban polygon [14]; background map [106]. Map realized with Esri ArcGIS 10.6.1 [48].

identified at coarser spatial scales, within countries at province level for example.

6.8   No dice rolling for target selection: the Iraqi example

As discussed in Section 2.4, we observed that terrorism may attack both civilians and non-civilians. The mechanisms involve in terrorism targeting may differ according to various factors. As a first exploratory step, terrorist attacks can be classified according to four classes associated with different reasons behind the choice made by terrorist groups to launch an attack. For example, Drake classifies targets into: symbolic, functional, logistical, and expressive targets [36, p. 9].

Terrorists may attack symbolic targets, primarily to hurt the psychological target—one or several individuals: a witness of a terrorist event, a radio audience, a family, or government representatives, for example—and they do so to get a reaction or overreaction from it that would justify further violence. Attacking symbolic targets attracts attention of the public [26, p. 1-2] and satisfies the desire of terrorist groups for renown and reaction [122, p. 131].

Terrorists may also attack functional targets, which are considered as a threat for terrorist groups. A functional target could be a traitor, a police officer, or a politician. Moreover, terrorists may attack logistical targets in order to optimize their resources. Logistical targets provide money, weapons and other useful elements to terrorists.

The classes of targets may overlap since the choice made by terrorist groups about their target is generally determined by multiple causes that may fall into more than one category. Moreover, the analysis of terrorist targets does not necessarily shed light on the real intention of the terrorists [36, p. 9]. However, classification of targets may help understand potential mechanisms behind the processes of target selection.

More generally, target selection is determined by various factors that include human, material, financial resources, and the leadership of the terrorist group [36, pp. 73-97]. A desire for revenge [122, p. 131], indignation, or anger may push individuals or groups to employ terrorism as an expressive target [36, pp. 9-15].

In addition, the ideology of a terrorist group may orientate its choices. Ideology may set the limit within which a group selects its targets [36, p. 16]. In order for terrorist groups to attain political goals [63, p. 40], an efficient strategy needs to be defined in order to achieve them [36, p. 35]. Therefore, one aspect of the strategy of terrorist groups is to maximize the impact on the psychological target [36, p. 53].

For the sake of illustration, we explore the frequency of targets classified into various categories based on data from terrorist attacks that occurred in Baghdad between 2002 and 2017. The results are displayed in Table 6.1.

In this case study, we observe that terrorist groups mainly targeted civilians. The most targeted groups are private citizens and property (3,243 attacks), followed by business (1,140 attacks). Non-civilians have not been spared, however. Governmental targets count a total of 872 attacks (3rd position), followed by police (682 attacks) and military (359 attacks) targets.

Table 6.1: Frequency of attacks by type of targets: Baghdad, Iraq (2002-2017). Note that the targets of 173 attacks are unknown (final row). Data source: GTD [56].

Type of target	Frequency
Private Citizens & Property	3242
Business	1140
Government (General)	872
Police	682
Military	359
Transportation	258
Religious Figures and Institutions	233
Terrorists and Non-State Militia	114
Educational Institution	90
Government (Diplomatic)	71
Journalists & Media	65
Utilities	37
Other	35
Airports & Aircraft	12
NGO	10
Violent Political Party	7
Food or Water Supply	4
Tourists	4
Telecommunication	3
Maritime	0
Unknown	173

This example based on the analysis of terrorist attacks in Baghdad from 2002 to 2017 shows the wide range of types of potential targets. Targets are selected by terrorist groups according to various factors. Since targets tend to be more similar at close proximity (Tobler's law) and terrorist groups tend to regularly attack specific targets, cluster patterns are often observed at various spatial scales.

The relative proportion of attacks per type of targets may considerably vary among cities, regions and countries and may change over time. Furthermore, one should keep in mind that overlap between classes used to define targets cannot be excluded and the definition of the classes remain subjective to a certain extent and may influence the results and their interpretation.

6.9 Conclusion: terrorism is clustered at various spatial scales

In this chapter, we have explored the spatial nature of terrorism using GTD. We indicated that terrorism may be perceived as random, but in practice, we often observe that terrorist groups tend to favor specific targeted locations and the resulting clustering patterns is observed at various temporal and spatial scales.

Terrorism, as most social phenomena, is affected by spatial autocorrelation, which makes its statistical analysis challenging. Standard statistical models cannot be applied without adjustment since their main assumptions are violated. Statistical tools, such as point processes, can capture and analyze the spatial structure associated with the spatial dependencies of terrorist events. This type of analysis may reveal the underlying processes behind the observed patterns.

Scholars have brought evidence that terrorism is clustered at various spatial scales. We showed that one of the world's most targeted country, Iraq, exhibited major clusters of extremely high intensity of terrorism. The cities of Mosul and Baghdad were strongly affected by terrorism between 2002 and 2017 in particular. Thus, we put emphasis that the detection of clusters and our understanding of the factors involve in clusters are dependent on the scale on which they are identified.

The interpretation of results from spatial analysis of terrorism data using GTD is restricted to the spatial accuracy of the localization of the events. Using the city of Baghdad as example, we showed that the database provides the exact same coordinates (city's centroid) for all attacks that occurred in the city, which does not reflect the true spatial distribution of the attacks in this city.

Given the relatively coarse spatial accuracy of terrorism data, one cannot capture within-city spatial heterogeneity of terrorist attacks using GTD alone. We suggested potential scenarios that could describe the true spatial distribution of the attacks in the urban area of Baghdad, based on the simulated patterns from three types of point processes. This illustrated example showed the limitations of the spatial accuracy of terrorist attacks from GTD.

Terrorists do not play dice in selecting their targets. Rather, they tend to specifically choose their targets, and hence, determine the locations of the attacks, at least to a certain extent. Terrorist targets may be classified as symbolic, psychological, logistic, or expressive targets. We illustrate the diversity of targets using attacks that occurred in the city of Baghdad between 2002 and 2017. In this case study, we showed that some targets are more targeted than others.

Terrorist attacks tend to target specific locations in order to maximize their chance to achieve their goals. The perceived spatial randomness of terrorism, as sometimes relayed by the media, does not appear supported by evidence, at least, at spatial scales where the locations of events can be considered accurate. Instead, we observe that terrorism tends to form clusters at various spatial scales and the processes behind the observed patterns can be analyzed through statistical methods.

So far, we have described clusters of terrorism as fixed areas in a study region. Yet the locations of clusters are not necessarily static over time. The number of terrorist events per spatial unit, the magnitude (number of deaths, number of casualties) of events, the number of spatial clusters, and the areas covered by spatial clusters of terrorism may vary over time. Next chapter will discuss in further details clustering processes of terrorism in both space and time.

Glossary

Clustered spatial point process: a clustered spatial point process assumes that the spatial points (events) attract each other. As a result, the realizations of the process (point pattern) exhibit spatial clusters [65, Ch.3] .

Complete spatial randomness: in the point process literature, a complete spatial random (CSR) process refers to a spatial process in which the observed points (events) do not stochastically depend from each other. In other words, there is no interaction between events: the location of an event does not influence (attract or repulse) other events. This process can be described by an (homogeneous) spatial Poisson point process [74].

First law of geography: the spatial dependency observed in most natural and social phenomena was well known by the geographer Tobler who stated that: "everything is similar, but near things tend to be more similar" [141]. His statement was so popular in the field of geography that it became later known as the "first law of geography".

Point pattern: a point pattern represents "a collection of points in some area or set and is typically interpreted as a sample from (or realization of) a point process" [65, p. 23]. In space, the set of spatial points that characterizes a spatial pattern and locations refers to events [33, p. 1].

Point process: a point process is defined as a "stochastic process whose realisations consist of countable set of points" [33, p. 199].

Poisson point process: an (homogeneous) Poisson point process is characterized by the following properties: the number of events of any disjoint regions are independent and follow a Poisson distribution with constant mean.

It results that events are equally likely to occur anywhere in the spatial domain [5]. Clustering and its antagonist, inhibition phenomena can be identified and quantified in terms of deviation from a CSR process [69, p. 289].

Regular spatial point process: a regular spatial point process assumes that the spatial points (events) repulse each other. This phenomenon refers to an inhibition process. As a result, the realization of the process (point pattern) exhibits regular patterns [65, Ch.3].

Spatial cluster: a spatial cluster of terrorism is often defined as an area where more than the average the number of terrorist events is observed [2].

The expected average of the number of events per spatial unit depends on the domain of investigation (e.g. within a city or country boundary) and the spatial level of aggregation (e.g. 5km grid-cell, province). Furthermore, the extent to which a value deviates from the mean is based on thresholds that are often subjective.

Hence, subjectivity with regard to the number and extent of spatial clusters identified in a study cannot be excluded. Therefore it is crucial that analysts are transparent and justify their choice with regard to the spatial domain and thresholds to define clusters .

FURTHER READING

Anselin, L. (1988). Do Spatial Effects Really Matter in Regression Analysis? *Papers in Regional Science,* 65(1): 11–34.

Anselin, L. (1990). Spatial dependence and spatial structural instability in applied regression analysis. *Journal of Regional Science,* 30(2): 185–207.

Diggle, P.J. (2014). Statistical Analysis of Spatial and Spatio-Temporal Point Patterns. *CRC Press,* Boca Raton, FL, USA, 3rd edition.

Drake, C.J.M (1998). Terrorists' Target. *St. Martin's Press,* St. Martin's Press,New York, NY, USA.

Hoffman, B. (2006). Inside Terrorism. *Columbia University Press,* New York, NY, USA.

Tobler, W.R. (1970). A computer movie simulating urban growth in the Detroit region. *Economic geography,* : 234–240.

CHAPTER 7

Myth No 6: Hotspots of Terrorism are Static

> Most of an organism, most
> of the time is developing
> from one pattern into
> another, rather than from
> homogeneity into a pattern.

> Alan Turing

7.1 Introduction: the dynamic nature of hotspots of terrorism

In the previous chapter, we showed that the location of terrorist events and their characteristics such as the number of deaths tend to aggregate in space at various spatial scales. This clustering phenomenon leads to local peaks that can be observed in both the number of events and deaths due to terrorism in specific areas.

However, local peaks of terrorism cannot be observed through data analysis due to the limitations of the spatial accuracy of terrorism data. We emulated potential spatial distributions of terrorist events within the city boundary of Baghdad that reflect hypothetical scenarios. We showed that a clustering process can be described and quantified as a deviation of a complete spatial randomness (CSR) process.

In practice, one can assess the level of spatial clustering, for a given time period, by counting the number of attacks per spatial unit across a spatial domain. For example, one may observe a total of 3,000 attacks in a country of 100,000 square kilometers. In this scenario, the country shows an average of 0.03 attacks (3,000 divided by 100,000) per square kilometer.

Hence, one can assess the local risk of terrorism based on how much the observed number of attacks per one square kilometer cell derives from the country average of 0.03 attacks per square kilometer. The same procedure can be replicated on the number of deaths per spatial unit (or any other metric of interest) to assess different components of the risk of terrorism at fine spatial scale.

Areas affected by abnormally high levels of terrorist activity (hotspots) can be detected at various spatial scales. The size and extent of the observed hotspots are typically affected by the choice of the spatial unit—in a regular lattice model, the spatial unit corresponds to the size of a grid-cell—and the choice of a threshold above which locations are defined as areas of high terrorism intensity, also called hotspots.

So far, we have considered hotspots as static spatial objects: we assumed that their number and shape does not change over time. However, this assumption may not reflect the dynamic nature of terrorism observed in various regions. In this chapter, we will explore the possibility of change over time of both the size and location of areas at risk of terrorism. Thus, the spatio-temporal analysis of local risk of terrorism will lead us to investigate *diffusion* processes which may occur in the neighborhood of local hotspots.

7.2 Contagious and non-contagious factors that cause the spread of terrorism

By analogy with the chemical process of diffusion in which a flow is generated from areas of higher concentration to areas of lower concentration (coldspots) [40], terrorism violence may spread from areas of higher terrorist activity to neighboring areas. The antagonist process, dissipation, also called negative diffusion, is associated with a decrease in terrorism activity in the neighborhood of areas of high terrorism activity.

Some authors distinguish diffusion from *contagion*: the former refers to the general process of movement, whereas the latter refers to one mechanism for achieving that movement [23]. This distinction opens the possibility for non-contagious mechanisms involved in the movement of a diffusion processes. The main principle of contagious and non-contagious diffusion processes is described in further details below.

A contagious diffusion involves two objects: (a) an event which may have potentially triggered the process of diffusion and (b) the locations affected by the presence of a triggering event [50]. By analogy with contagious disease's spread, contagious diffusion depends on direct contact between events [23]. In this framework, spatial proximity is a necessary condition.

Diffusion that does not require direct contact refers to non-contagious diffusion, which can be triggered by spontaneous innovation or imitation [23]. An increase in the intensity of terrorist events observed at the proximity of terrorism hotspot is not necessarily due to spatial proximity with other terrorist events. Rather, it could be the result of common characteristics (e.g. high density of symbolic targets) shared between the hotspot and its neighborhood [93, 62, 102].

There are numerous potential non-contagious factors of terrorism diffusion. They include: modern mass media ([148, p. 103], [30]), financial support [62], transnational collaboration ([62], [28, p. 17]), freedom of movement and transportation ([62], [148, p. 189]), state capacity ([61, 55, 153]), state policy [97, p. 49], population density ([29, p. 115]), and the dynamics of the terrorist group (e.g. tactical change) [117, 81].

In particular, the mass media [148, 89, 30] or transnational collaboration [62, 28] may encourage terrorist groups active in the vicinity of a cluster to imitate or reproduce the attacks perpetrated in the cluster that have been relayed by the media. In this case, the causes of diffusion may differ from those involved in a contagious diffusion context. For example, tactical information or money shared via Internet among terrorist groups that act in neighboring areas are not affected by spatial distance—electronic correspondence and exchanges among terrorist actors are almost instantaneous even at very large distance.

Terrorist groups may also decide to act far beyond their usual areas of operations and their close neighborhood. Terrorist groups may be forced to change their main target locations due to counterterrorist actions that threaten their activity. In addition, terrorist groups change their area of activity for tactical or strategic advantages. Based on a study of civil war, Schutte and Weidmann describe this phenomenon as *relocation* diffusion, in which terrorist groups cease to target a specific location and relocate their activity beyond the close neighborhood [131].

Since relocation diffusion can theoretically occur at any distance from the initial areas of high terrorism activity, its identification remains challenging. Locations that encounter relocation diffusion might be confused with an onset of terrorist activity that may occur independently from the initial location of the source of terrorism. For the sake of clarity, we will not consider relocation diffusion in this study.

7.3 Type of terrorism diffusion is associated with tactical choice

Diffusion processes in terrorism have been traditionally analyzed in the temporal dimension only [59, 54, 42, 111, 117]. These studies have pointed out how high activity of terrorism in a specific time period may increase terrorism activity in future time periods.

Few studies have investigated diffusion processes in space and time [91, 52, 131, 88, 85]. These studies have brought evidence that high activity of terrorism in specific areas can be associated with an increase in terrorism activity in neighboring locations.

Measuring diffusion has proved to be challenging. Forsberg considers diffusion as unobservable because it is usually derived from measures of correlation or proximity in space or time between events [50], which can be inaccurate, as reported by Black [10]. However, recent approaches have improved the accuracy of measures of diffusion processes and in particular their uncertainty [152, 151].

Empirical studies showed that both contagious and non-contagious processes might operate simultaneously. Based on the analysis of terrorism that occurred in three couples of neighboring states (Lebanon-Israel, Colombia-Peru, and India-Pakistan), Cliff and First found that both non-contagious and contagion diffusion can occur across borders [21]. Moreover, the authors found evidence that contagious and non-contagious diffusions reflect different tactics employed by terrorists.

LaFree et al. compared terrorist events perpetrated by the Spanish terrorist group Euskadi Ta Askatasuna (ETA) in French and Spanish regions between different time periods [81]. They identified and quantified the differences in spatial patterns of terrorist attacks over time. Similar to Cliff et al., they suggested that contagious and non-contagious diffusion reflect different strategies adopted by ETA.

Empirical findings carried out at local level showed that diffusion of terrorism occurs mainly in failed states. Failed states are often viewed as states lacking of financial, human resources, and infrastructure to control their territory [124, p. 70]. Therefore, they encounter difficulty to block terrorism that spread from hotspots within their territory or diffusion across border generated by hotspots located in close proximity with neighboring countries [55, 61, 153].

In the next section, we will investigate changes in the terrorist activity from the Islamic State (ISIS) in a typical failed state: Iraq. We will highlight areas of high terrorism activity and illustrate diffusion processes that may occur in the neighborhood of hotspots of terrorism.

7.4 Scale and magnitude of the clustering process associated with ISIS attacks perpetrated in Iraq (2017)

ISIS was the most active terrorist group in the world in 2017. However, over the year, the group's activity has declined in various parts across Iraq. In this section, we will describe and quantify changes in the group's activity through the analysis of the spatial patterns of their attacks. This work will be carried through a point process approach.

A point process approach has several advantages. First, it does not require us to aggregate the data—which would be necessary within a lattice framework. For example, one would aggregate the data into each province in Iraq. The aggregation procedure would ineluctably result into a loss of information on the spatial coordinates of the individual attacks—the longitude and latitude of events in a given province are given the same pair of coordinates (usually the centroid of the province) which results into a loss of spatial accuracy.

Second, a point process approach allows us to make inference on the underlying processes that have generated the observed point patterns. This means that we will assess the effects of factors on the intensity of the point process to better understand what drives terrorism to form clusters.

In the previous chapter, we have described and compared different point processes (random, uniform, or clustered). We will refer to these processes to quantify the degree of the deviation of the observed patterns from a CSR process, and hence, estimate the magnitude and extent of the clustering process.

Our first exploratory work consists of determining if ISIS attacks are clustered over the entire study period, from January to December 2017. In this initial phase, we do not consider changes over time to first get a general overview of the scale and magnitude of the clustering process of ISIS attacks in Iraq before investigating its dyanamics.

Measuring the range of scales and the magnitude of clustering of terrorist attacks provides insight into the organization and strike capacity of the perpetrator of the attacks. Clustering within a short radius indicates that terrorist groups target locations very close to each other. A clustering of similar magnitude but extended to a larger radius would suggest a reinforcement of the group's activity beyond their usual area of activity, which may be associated with higher resources and organizational capacity.

In order to identify and quantify the scale and magnitude of spatial clusters, we compute the pair correlation function (pcf) for various radii (in degree) and report the values on the y-axis corresponding to each radius r on the x-axis in Figure 7.1. The pcf is a function that computes the ratio between the probability of observing a pair of points separated by a distance r and the corresponding probability values for a CSR process.

Values of the pcf (*plain line*) above 1—the value 1 is the expected pcf value for a CSR process (*dashed line*)—indicate a clustering process. The higher the value, the more magnitude of the clustering. Values below 1 suggest inhibition or regularity (see Section 6.7 for a description of different point processes). Note that the values of the pcf are adjusted to reduce bias due to edge effects using Ripley's isotropic edge correction (for further details on edge effects and method to mitigate them, see [5]).

Figure 7.1 shows that terrorist events are clustered at fine spatial scale until a radius of about 0.5 degree (approximately 50 km) above which the pcf takes values close to 1. The results suggest that ISIS attacks are highly clustered at very close proximity and the clustering process weakens until about 50 km.

Beyond approximately 50 km, the observed point pattern cannot be dissociated from a realization of a CSR process. Consequently, ISIS attacks may be considered spatially independent above this distance. In practice, this means that areas very close to regions affected by terrorism are more likely to encounter terrorist attacks. However, the risk of terrorism in a given location appears not influenced by attacks distanced of more than 50 km.

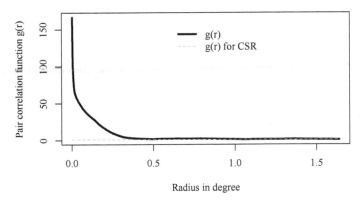

PAIR CORRELATION FUNCTION
ISIS TERRORIST EVENTS IN IRAQ (2017)

Figure 7.1: Pair correlation function (pcf) based on the observed ISIS terrorist events that occurred in Iraq in year 2017. The *plain line* shows the pcf function $g(r)$ computed for various radius values (r): from 0 to 0.6 degrees (about 66 km). The pcf line $g(r)$ *for CSR* for complete spatial randomness (CSR) is equal to 1 (*dashed line*). CSR corresponds to an homogeneous Poisson process. Data source: GTD [56].

Various reasons can lead to the fine-scale clusters observed at distances below 50km. Spatial proximity with terrorist events (contagious factor) and non-contagious factors such as shared characteristics of neighboring targets can explain the observed clusters. Areas highly affected by terrorism may count a high density of symbolic targets, be densely populated, and therefore, the attacks may target a high number of potential victims with an increased propensity to be publicized.

7.5 Localizing and quantifying the reduction of ISIS activity from January to December 2017

In the previous section, we showed that in Iraq in 2017, ISIS attacks clustered at fine spatial scales until about 50 km, above which one cannot exclude that events are spatially independent. This preliminary work considered all events that occurred in the time period without accounting for potential changes over the year.

We learned from Section 5.7 that ISIS activity reduced in Iraq from mid-2017 until the end of 2017. The end of 2017 corresponds to the collapse (or apparent collapse) of ISIS and the end of the civil war [7, 149]. One may want to know when was the strongest reduction of ISIS activity, which could be associated with important events in the country. Also, one could ask how a reduction of ISIS activity may affect the magnitude and the extent of the clustering process.

Below, we investigate monthly changes in the magnitude and scales of the clustering patterns associated with ISIS terrorist attacks in Iraq. For this purpose, we estimate and compare the values of the pcf at various radii, this for each month over 2017, and report the values in Figure 7.2.

To ease the reading of the plot, we use four different colors which distinguish each trimester of 2017 (Q1: Jan to Mar (*purple*); Q2: Apr to Jun (*blue*); Q3: Jul to Sep (*green*); Q4: Oct to Dec (*yellow*)). For each trimester, we dissociate each month using three line types (1st: *plain*; 2nd: *dashed*; 3rd: *dotted*).

The magnitude of the clustering process is indicated by the values of the pcf. We observe that in most months the pcf decreases faster in Q3 and Q4 compared to Q1 and Q2. The smaller range of clustering scales observed in the second half of the year suggests a weakening of the terrorist activity. The clustering extent is reduced so that terrorist events can be considered spatially independent already above 0.2 degree, which corresponds to approximately 22 km.

So far, we have observed and quantified a reduction of the clustering process over time, this, for various radii between pairs of points. A complementary analysis is to determine where did the changes of terrorist activity occur, which is crucial information for counterterrorism actors that need information on the spatial locations that encounter reduction of terrorism activity.

A simple descriptive approach consists of visualizing and comparing for each month the locations (geographic coordinates) of terrorist attacks on a two-dimensional map. Figure 7.3 shows the location of terrorist events in Iraq, this for each month from January to December 2017.

Figure 7.3 shows that Mosul (*black star, top*) and Baghdad (*black star, center*) are highly targeted by terrorism in the first months (JAN-JUL), followed by a reduction of the attacks towards the end of the year. This observed decrease in the number of attacks in these two cities is generalized to many regions across the country, as illustrated by a decrease in the number of *red crosses* in most provinces in Iraq in the second half of the year in particular.

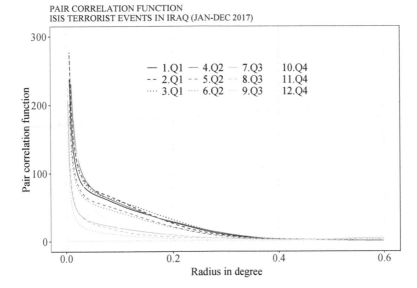

PAIR CORRELATION FUNCTION
ISIS TERRORIST EVENTS IN IRAQ (JAN-DEC 2017)

— 1.Q1	— 4.Q2	— 7.Q3	10.Q4
-- 2.Q1	-- 5.Q2	-- 8.Q3	11.Q4
···· 3.Q1	···· 6.Q2	···· 9.Q3	12.Q4

Figure 7.2: Pair correlation function (pcf) based on the monthly observed ISIS terrorist events that occurred in Iraq from January to December 2017. The pcf function is computed for various radius values: from 0 to 0.6 degrees (about 66 km) for each month (1st: *plain*; 2nd: *dashed*; 3rd: *dotted*) within each trimester (Q1: Jan to Mar (*purple*); Q2: Apr to Jun (*blue*); Q3: Jul to Sep (*green*); Q4: Oct to Dec (*yellow*)). The *gray dotted line* has value 1, which corresponds to the pcf for an homogeneous Poisson process (complete spatial randomness). Data source: GTD [56].

7.6 Explaining and visualizing diffusion of ISIS activity from January to December 2017

A visual inspection of changes in the number and location of terrorist attacks is instructive. However, it does not suffice to reveal the mechanisms behind the changes in the observed spatial patterns. We will use a point process modeling framework to identify and quantify the factors associated with its spatial intensity (the expected number of points in a given area), identify areas of abnormally high intensity of terrorism, and quantify diffusion processes that occur in areas at risk.

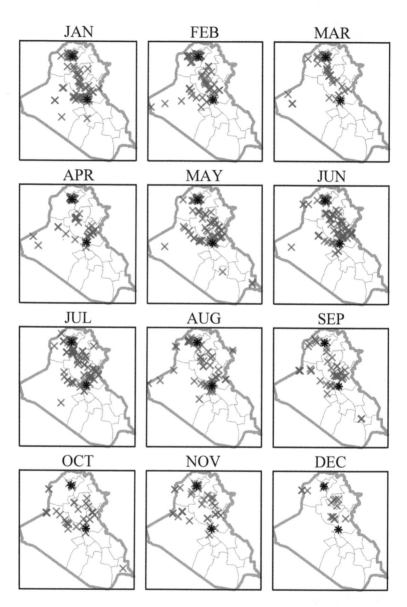

Figure 7.3: **Observed ISIS terrorist events in Iraq (Jan-Dec) 2017.** The maps highlight the location of terrorist attacks (*red crosses*) perpetrated by the Islamic State (ISIS) for each month from January to December 2017. The centroids of Mosul and Baghdad are represented by a *black star* in the top and center of the map, respectively. Data source: GTD [56].

We use a log-Gaussian Cox process (LGCP) (see Shortbox) [69, p. 291] to model the spatial distribution of terrorist attacks perpetrated by ISIS from January to December 2017. The LGCP belongs to the class of stochastic models. This class of models has the property to account for randomness. These models are described in further detail in the next chapter (Section 8.2). The LGCP is a point process model that accounts for the spatial dependency present in the data and allows us to analyze the effects of environmental and socio-economic factors on the (log) intensity of the process [6, pp. 355-356].

The strike capacity of a terrorist group can be affected by the terrain conditions and their strategies may be conditioned to the number and accessibility to the targets. Therefore, we include altitude and population to account for this. ISIS may cooperate with some terrorist groups, and join their efforts to target similar targets and locations, or in contrast, ISIS may avoid areas affected by adversary terrorist groups. We therefore include travel time to the nearest non-ISIS attack and a count of number of non-ISIS attacks.

The results suggest that the (log) intensity of terrorist attacks increases in regions of lower altitude. The world's two most targeted cities, Baghdad (34 m.) and Mosul (223 m.), are indeed large cities located in relatively low altitude, which can therefore be easily accessed by ISIS and offer a large potential number of targets.

Furthermore, the results indicate that ISIS attacks are spatially more clustered close to areas targeted by terrorist groups other than ISIS and in locations with higher number of attacks perpetrated by other terrorist groups. These latter results may suggest positive interactions, and therefore potential cooperation, between ISIS and other terrorist groups active in Iraq.

In order to account for the fact that the intensity of terrorism may vary over time between January to December 2017 and can be influenced by previous attacks, we allow the intensity of the LGCP to be temporally correlated. We use a first-order temporal autocorrelation process AR(1) model, where the intensity in a given month depends on the intensity of the previous month. This approach expresses a simple departure from independence in time series [31, p. 19].

For each grid-cell (approximately 4km resolution), we define areas of high terrorism intensity, as locations where the values of the point process intensity is higher than the 95th percentile measured within the spatio-temporal domain. Hence, we observe change in the expected intensity between successive months from January to December 2017.

We illustrate the results of the analysis in two regions that have been affected by important changes in terrorism activity. Figure 7.4 focuses on three northern provinces in Iraq, which include Ninawa (Mosul), Dihok, and Arbil. Figure 7.5 highlights the region of Baghdad along with Diyala province, a neighboring province in the northeast of the capital of Iraq.

From January to December 2017, ISIS showed a decrease in terrorist activity in most Iraqi regions, which includes Mosul (Figure 7.4) and Baghdad (Figure 7.5) areas. High levels of terrorism activity are observed from April to June in both Mosul and Baghdad areas, with an increase in the sizes of hotspots (*green* polygons), which can be associated with a diffusion process. In fall and winter, the activity is reduced which translates into a reduction of the number and the size of hotspots (*green* polygons) over the year, which corresponds to a dissipation process.

While ISIS was a major terrorist threat during the Iraqi civil war (2014-2017) [49, p. 27], the group's activity has been drastically reduced in the second half of the year 2017. The dynamic patterns of ISIS attacks captured by the LGCP reflect reduction in the levels of their activity that might have resulted from several possible mechanisms such as, e.g., increase in counterterrorism of activity and/or changes in the terrorist strategy or resources and/or competition among terrorist groups, which may depend or not to spatial proximity.

7.7 Conclusion: change is the only constant in terrorism

In this chapter, we have investigated the dynamic nature of terrorism activity which is reflected into changes in the observed patterns of terrorist events.

We showed that spatial data, including terrorist events, can be viewed through different lenses. One may make inference at subnational levels of analysis through a lattice approach that assumes terrorism aggregated within spatial polygons. Within a geostatistical framework, one may be interested in modeling a characteristic of terrorism, assumed to be continuous in space, which is measured at the location of events.

As an alternative, we opt for an approach that allowed us to capture the scale and the extent of the clustering process of terrorist events. Our case study showed that attacks from the Islamic

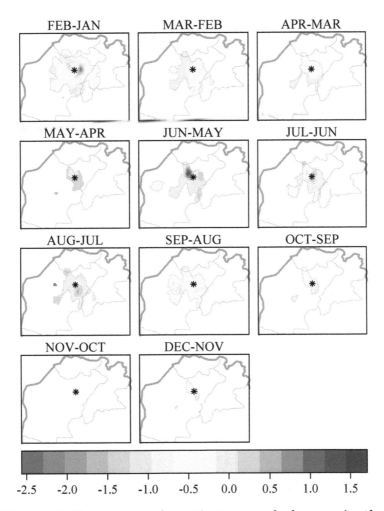

Figure 7.4: **Hotspots and spatio-temporal changes in the intensity of terrorism in Ninawa (Mosul), Dihok, and Arbil provinces (*thin gray lines*) in Iraq (*thick gray line*) from January to December 2017).** Pixel-level (4 km) monthly differences between two successive months in the expected intensity of a log-Gaussian Cox spatio-temporal point process in hotspots (*green* polygons). Monthly decrease (negative values: *blue*) and increase (positive values: *red*) in terrorist activity. The centroid of Mosul is represented by a *black star*. Data source: GTD [56].

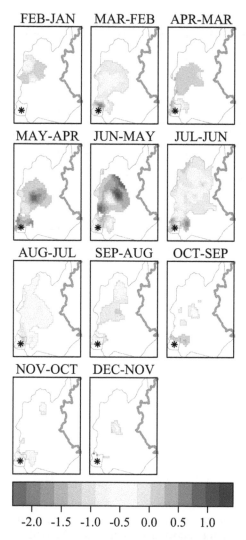

Figure 7.5: **Local hotspots and spatio-temporal changes in the intensity of terrorism in Baghdad and Diyala provinces (*thin gray lines*) in Iraq (*thick gray line*) from January to December 2017.** Pixel-level (4 km) differences between two successive months of the expected intensity of a log-Gaussian Cox spatio-temporal point process in hotspots (*green* polygons). Monthly decrease (negative values: *blue*) and increase (positive values: *red*) in terrorist activity. The centroid of Baghdad is represented by a *black star*. Data source: GTD [56].

State (ISIS) in Iraq clustered in 2017 at various spatial scales and the clustering process has been weakened in second half of the year.

This illustrative example, based on a point process model, allowed us to better understand the spatio-temporal processes of terrorist attacks perpetrated by ISIS in Iraq at a monthly level, this over the year 2017. Our method captured the underlying processes behind the observed patterns of terrorism, and associated them with geographical and socio-economic factors.

In Mosul and Baghdad areas, we observed an extension of hotspot areas in summer and a drastic shrinkage of terrorist activities in the second half of 2017. These changes in terrorism activity might be the results of interactions between ISIS, other terrorist groups, and counterterrorism forces.

While some regions are consistently affected by a relatively high level of terrorism, we observed that at fine spatial scales, hotspots are not static over time. Even in areas often extremely affected by terrorism such as Mosul and Baghdad in Iraq, terrorism activity may exhibit high variations in both space and time.

Analyzing clusters through the Log-Gaussian Cox process

From a spatial point process viewpoint, a clustered spatial distribution of terrorist events—the observed terrorist events represent a realization of a point process—may be the result of an underlying process that can be described by a Cox processes, named after the statistician David Cox, who first published the model in 1955.

Cox processes, also called doubly stochastic Poisson process, are an extension of the inhomogeneous Poisson process; the latter allows for spatial variation of the expected intensity, as a function of space, which is itself an extension of the homogeneous Poisson process that assumes a fixed expected intensity with no interaction among points (CSR).

Within a Cox process framework, the intensity function is a realization of a random field, which allows for overdispersion in space [69, p. 291]. Covariates along with unobserved effects (random terms) can be used to explain variations in the random intensity. The Log-Gaussian Cox process (LGCP) is a subclass of Cox processes, where the logarithm of the intensity is a Gaussian random field, and therefore analytically tractable [6, pp. 355-356].

Glossary

Autoregressive models AR(1): express a departure from independence in time series [31, p. 19]. They may be formulated as a first-order difference equation [58, p. 1]:

$$Y_t = c + \rho Y_{t-1} + \epsilon_t \qquad (7.1)$$

where c is a constant, the dependent variable (Y_t) takes value at discrete time period (t) and is related only to its previous value (Y_{t-1}) through a parameter measuring autocorrelation (ρ) and an error term (ϵ_t), the latter consists of serially uncorrelated random variables with zero mean and finite variance [58, p. 7]

Contagious diffusion: diffusion process that requires spatial proximity (direct contact) with areas that exhibit higher than average concentration of terrorist events (hotspots) [23].

Diffusion: from Latin (*diffundere*): to pour out or shedding forth [107]. By analogy with the chemical process of diffusion in which a flow is generated from areas of higher concentration to areas of lower concentration (coldspots) [40], terrorism violence may spread from areas that exhibit higher than average concentration of terrorist events (hotspots) to neighboring areas.

Dissipation: also referred to as negative diffusion. A dissipation process of terrorism represents a decrease in terrorism activity in areas that previously exhibited higher than average concentration of terrorist events (hotspots).

Failed state: a failed state can be defined as a state that exhibits "inability or unwillingness to protect its citizens from violence and perhaps even destruction" [18, pp. 1-2].

Non-contagious diffusion: diffusion process that does not require spatial proximity with areas highly targeted by terrorism. This includes terrorist events that would occur, e.g., by an imitation process independently of the location of hotspots [23, 50].

Relocation diffusion: process where the spatial location of hotspots disappears and relocates beyond the close neighborhood. This could occur, e.g., when a terrorist group losses access to its favored target location due to counterterrorism activity.

FURTHER READING

Crenshaw, M. (1981). The Causes of Terrorism. *Comparative Politics*, 13: 379-399.

Forsberg, E. (2014). Diffusion in the Study of Civil Wars: A Cautionary Tale. *International Studies Review*, 16: 188-198.

Illian, J. and Penttinen, A. and Stoyan, H. and Stoyan, D. (2008). Statistical Analysis and Modelling of Spatial Point patterns. *John Wiley & Sons Ltd*, Chichester, West Sussex, UK.

Zammit-Mangion, A. and Dewar M. and Kadirkamanathand, V. and Sanguinetti, G. (2012). Point Process Modelling of The Afghan War Diary. *PNAS*, 109: 12414–12419.

CHAPTER 8

Myth No 7: Terrorism Cannot be Predicted

> Those who have knowledge, do not predict. Those who predict, do not have knowledge.
>
> Lao Tzu

8.1 Prediction of terrorism: statistical point of view

One may think that since terrorism is a highly complex social phenomenon that does not abide by social laws, it cannot be predicted through the use of quantitative models. A similar statement can be also found in the following claim: "acts of terrorism cannot be summarized by mathematical formulas".

Various reasons may lead to the conclusion that terrorism is inherently unpredictable. First, one may assume that the realms of social science differ from those of natural science (ontology) and one cannot access knowledge (epistemology) of social phenomena through quantitative methods. Within this framework, the mathematical language and its derived tools such as predictive statistical models are rejected.

Second, the role of statistical explanatory and predictive models is often misunderstood. This can be the result of a lack of sufficient training in statistics, which generates confusion on the usage, limits, and interpretation of predictive models applied in social science and other research areas.

Third, outside the statistical community, the concept of prediction is sometimes assimilated to prophecy, or reduced to a deterministic statement about a future state. Such confusion around its concept can be misleading.

As a result, the role of predictive models in social science has been under controversy. In this chapter, we opt for clarifying a few statistical concepts in the hope that the non-specialist reader can get familiarized with fundamental notions required to better understand and appreciate the meaning and implications of statistical models applied to predict terrorism.

8.2 Stochastic models for the statistical prediction of terrorism patterns

Systems described through mathematical concepts and language— these systems often refer to the so-called mathematical models— have a long tradition in the study of political conflict. Lichbach identified more than 200 scholarly works which use mathematical models to describe phenomena such as guerrilla wars and insurrections [86].

Mathematical models can be further distinguished into *deterministic* and *stochastic* models [31, p. 59]. A well-known set of deterministic models are Newton's three laws of motion and the law of universal gravitation described in his famous work *Philosophiae Naturalis Principia Mathematica* first published in 1687 [100, p. 194].

In Newton's model, the position of the Earth around the Sun (Figure 8.1, *left*) is entirely determined and totally predictable if one ignores interactions with the other objects of the solar system. Note that the use of deterministic models does not necessarily imply that the processes under study are predictable. This is the case if one studies the trajectories of more than two bodies (e.g. the Sun, the Earth, and the Moon) through classical mechanics. For most initial conditions the system follows a chaotic process [73, p. 65].

Despite their immense scientific contribution, deterministic models alone are rarely suitable to predict social phenomena and terrorism is no exception. We understand the concept of prediction as "the process of applying a statistical model or [data mining] algorithm to data for the purpose of predicting new or future observations" [133]. You may observe that prediction does not only refer to temporal prediction (also called forecast) but also applies to, e.g., spatial prediction, when one makes prediction at locations that do not have data.

The causes of terrorism are multidimensional and operate at the individual, group, sub-national, national, and transnational levels [77, 122]. Ideology and belief are crucial factors of terrorism [28, p. 29], which may vary among individuals and groups of terrorists, and change over time.

Therefore, it is virtually impossible to predict terrorist attacks individually at any desired level of accuracy in space (exact coordinates) and/or time (exact time). Such deterministic prediction would imply that all processes involved in each terrorist event are known and entirely determined. The use of foreknowledge to forecast individual crimes, as portrayed in the movie *Minority Report* (2002) from Steven Spielberg belongs and is likely to remain in the realm of science fiction.

Not all predictive models are deterministic, however. While deterministic models predict a unique outcome from a given set of circumstances, stochastic models predict a *set* of possible outcomes through a probabilistic statement [139, p. 2].

The "electron cloud model" proposed by Schrödinger in 1926 [130] is a well-known stochastic model of the structure of the atoms. In this framework, the position of the electron surrounding the nucleus of a ground state hydrogen atom (Figure 8.1, *right*) is not known before it is measured. However, one can attribute a *probability* to identify the electron in a so-called "electron cloud" surrounding the nucleus.

Stochastic models have an interesting property, which is well described in the following anecdotal example. In 1777, Georges-Louis Leclerc, Comte de Buffon, was asked to find the probability that a needle will land on a line, given a floor with equally spaced parallel lines.

The movement of each needle is very complex, and hence, the exact position on the floor of each individual needle is hardly predictable. However, Buffon discovered that on a long-run, when a very large number of needles are considered, a fascinating pattern appear: the probability that a (short) needle will cross a line is exactly $\frac{2}{\pi}$ [146].

Analogously, characteristics of individual terrorist attack such as their geolocalization, time, or number of deaths cannot be exactly known before they occur due to the inherent stochastic nature of terrorist events. When taken individually they become unpredictable.

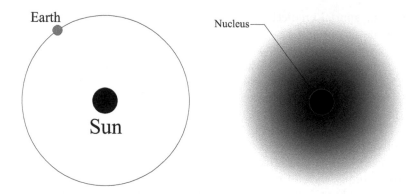

Figure 8.1: Schematic view of a *deterministic* and *stochastic* model (not to scale). *Left*: *deterministic* model of the orbit of the Earth (*gray disk*) around the Sun (*black disk*). In this two-object model, the position of the Earth is known with absolute certainty at any time of the year. *Right*: *stochastic* model of the position of the electron, represented by an "electron cloud" (*shaded area*) surrounding the nucleus (*black disk*) of a ground state hydrogen atom, with higher probability (*dark color*) close to the nucleus and lower probability (*bright color*) at farther distances.

As the needles, the unpredictability of individual terrorist acts does not imply that terrorism cannot be predicted at an aggregated level of analysis. When observed as a whole, one may capture, analyze and predict terrorism patterns using stochastic models—partially, a consequence of the central limit theorem (see Shortbox). Stochastic models have been successfully applied to detect trends and patterns of insurgent movements at fine spatial and temporal scales [150].

8.3 Predicting terrorism: limitations, opportunities, and research direction

Predictive techniques have been used to explore various facets of terrorism. To get an overview of the direction of research on terrorism prediction, a relatively simple approach is to extract and analyze the frequency of words used in the abstracts of academic articles that contain the words "terrorism" and "prediction".

Using an adapted version of the R code from `https://github.com/christopherBelter/scopusAPI`, we identify research articles from all scholarly journals indexed by Scopus, whose abstract contains the words "terrorism" and "prediction".

We note that a relatively small number of research works—944 articles in total—contain the words "terrorism" and "prediction" in their abstract. This indicates that little research with predictive purposes has been made in the field of terrorism so far. Indeed, predictive approaches in similar research areas have mainly focused on armed conflict, war, and to a lesser extent, insurgency events [87, 152, 143].

Second, we extract the most frequent words from a total of 944 abstracts to learn about potential topics of research related to the prediction of terrorism. A word cloud (Figure 8.2) highlights the most frequent words (words that appear more than 100 times) among all the recorded 944 abstracts. Words with a larger font size are those that are more frequent.

From visual inspection, we observe that among the most frequent ones stand the words "social", "research", "risk", and "model". This suggests that work on terrorism prediction can be associated with social research work and risk models. Predictive models are indeed often used to quantify risk.

Note that this exploratory exercise does not replace a rigorous meta-analysis which would be based on a thorough literature review on the topic. However, it provides us with an overview of possible directions of research using predictive techniques applied to terrorism studies.

In a related research area, Cederman and his colleagues carried out a meta-analysis that reviews predictive models applied to conflict events [13]. Their conclusions apply to terrorism research, which faces similar challenges.

First, the authors observed that so far, predictive research work has not been accurate enough to have an impact on policy making. Second, they indicate that there is a compromise between explanation and predictive purposes and put emphasis on the fact that explanation is more urgent than prediction to help policy makers prevent conflict.

Third, they mentioned that long-term and global predictions are more challenging than short-term and local predictions. Nevertheless, they consider the merits of predictive models to generate potential scenarios, if data are not too sparse.

In sum, three criteria are required so that predictions that can help policy makers designing and assessing targeted counterterrorism measures require: (i) quality of the data, which also includes its accuracy, (ii) quantity of data (sparse data makes predictions

Figure 8.2: Word cloud that represents the most frequent words (words that appear more than 100 times) from a Scopus research based on the words "terrorism" and "prediction" from the abstracts of 944 recorded articles before 2019. Words with a larger font size are more frequent. Data source: Scopus.

very challenging), and (iii) theoretical knowledge to understand the mechanisms behind the observed data [57, 13].

8.4   Artificial intelligence to serve counterterrorism?

Artificial intelligence (AI) can be defined as the ability of a machine to "reason, discover meaning, generalize, or learn from past experience" [24]. Since the last two decades, AI has progressively reshaped our society. The 29th March 2019 summit of the G7 science academies in Paris acknowledged the societal benefits of AI while pointing out potential risks of AI especially with regard to ethics.

The transformation of our society has been the results of growing development of a wide range of machine Learning (ML) algorithms. ML algorithms are a subset of AI in which algorithms are learning patterns often with the use of very large and complex data. ML algorithms have become ubiquitous in our day-to-day life. This includes voice and facial recognition systems for smart phone, or banking credit rating algorithms to name but few.

ML algorithms have been used to help counterterrorism forces identify terrorist suspects in airport based on sophisticated image and voice pattern recognition algorithms. Furthermore, some law enforcement agencies are already equipped with ML algorithms used to analyze network of criminal groups, detect and predict hotspots of crime [90].

In order to successfully learn, ML algorithms require a lot of data. The first known electronic database on terrorism counted 507 international terrorist events recorded from 1968 to 1974 by Jenkins in 1975 [70]. The relatively low number of observations would not have sufficed for typical ML algorithms to learn from the data. Furthermore, researchers implemeting ML algorithms did not benefit from the computational power of today's computers.

Consequently, terrorism scholars started using relatively simple statistical tools to analyze Jenkins' database. In 1980, Gleason showed that terrorism events are unlikely to be independent in time [54]. His early work opened the way to the quantitative analysis of terrorism.

The ongoing development of larger and more accurate terrorism databases along with sophistication of statistical models has led to more advanced methods to the analysis of terrorism in time [59, 111, 41, 45, 43, 42, 117]. Later, quantitative work extended to the analysis of spatial and spatio-temporal patterns of terrorism [82, 102, 44, 52, 81, 9, 104, 109, 94].

The development and application of statistical models in the field of terrorism has been essentially driven towards explaining rather than predicting terrorism. Predictive models using ML algorithms have been overlooked so far. One exception consists of few research works, that includes an attempt to predict terrorism worldwide at fine spatial scale a year ahead [34].

Despite their merits, the ML algorithms used to predict terrorism a year ahead are based on a yearly temporal aggregation, which remains too coarse to provide relevant information to counterterrorism forces active on the ground. Local policy makers require both fine spatial and temporal information to get tactical advantages over the perpetrators of the attacks.

With large datasets of sufficient accuracy and a choice of explanatory variables (also called "features" in ML jargon) selected based on theoretical knowledge, current ML algorithms can potentially deliver fine-scale predictions in both space and time and relevant for policy makers. In the next section, we illustrate the results of a ML algorithm trained to predict terrorism at fine spatial and a weak ahead in some regions highly affected by terrorism.

8.5 Machine learning algorithms to predict terrorism in space and time: a case study

In this section, we describe an approach aiming at predicting terrorist events at both fine spatial and temporal scales. In order to get enough data (on both presence and absence of terrorism in space and time), we constrain our study area in countries strongly affected by terrorism: Iraq, Afghanistan, and Pakistan with a temporal period covering 2002 to 2017. Although not being strongly affected by terrorism, we also include Iran since it lies between Iraq, Afghanistan, and Pakistan.

We use GTD to gather the spatial coordinates and the time of terrorist attacks that occurred from 2002 to 2017 in the study area [56]. Terrorist events are aggregated within a radius of 0.4 degree, which accounts for the uncertainty in the geolocation of GTD events [56]. The study area is discretized using a Voronoi diagram, named after Russian mathematician Georgy Feodosievych Voronoy. We split the study area into 706 polygons (Figure 8.3) so that for each city affected by terrorism, there is a corresponding region consisting of all points of the plane closer to that city than to any other.

We include various features that may be associated with the presence of terrorism. The choice of the features is crucial to make reliable predictions. It is mainly based on the work of Python et al. [114, 115] and described in further details below.

We consider the Gridded Population of the World (GPW v4.10) [98] and travel time to cities of more than 50,000 inhabitants [145] to capture potential targets, terrorism audience, and their access. Also, areas with high human activity tend to provide more potential targets. Other socioeconomic factors, such as the presence of armed conflict contributes to increase the risk of terrorism [114].

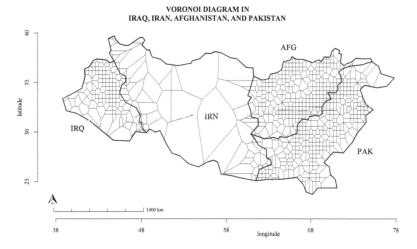

Figure 8.3: Predicting terrorism in space and time. The study area is discretized into a Voronoi diagram that covers Iraq, Iran, Afghanistan, and Pakistan. Small Voronoi polygons cover highly targeted areas while large Voronoi polygons cover less targeted areas. Data source: GTD [56].

We use NOAA satellite night lights [103], a widely used proxy for human activity [138, 39]. We extract conflict data from PRIO-GRID, which is a 0.5-degree grid that provides various geographic and socioeconomic variables worldwide [142].

Some terrorist groups, for example, the Taliban in Afghanistan may hide in mountainous areas which provide tactical advantages over governmental forces. Furthermore, oil fields are critical infrastructures that can be at risk of being targeted by terrorist groups. We capture these two factors by using the proportion of mountain and the presence of oil fields, respectively, which are provided by a database PRIO-GRID.

Furthermore, terrorist groups may use borders as safe havens to exchange information, money, and weapons. Given the logistic advantages, terrorist groups can therefore perpetrate attacks close to borders. This is captured as distance to national border data using PRIO-GRID [142].

The link between democracy and terrorism is not straightforward. Also, democracy may indirectly impact terrorism via its role on the media. Lower levels of democracy might reduce the rights of the press, which could, in turn, lead to a reduction of the number of reported terrorist attacks. We control for the level of democracy with a variable from PRIO-GRID [142] that measures democratic levels for each country.

Ethnically fractionalized areas [95, 121] may be more prone to conflict including terrorism. Terrorist groups may deliberately target ethnically fractionalized areas that may result into a disproportionate response from the government, which in turn, may increase support from the population towards the terrorist groups. We account for this using the variable ethnic fractionalization provided at high spatial resolution in [114].

Having prepared the data, we use XGBoost, which is a fast and efficient version of a gradient boosting class of machine learning algorithm [15].

The aim of the suggested approach is straightforward. We predict if one or more terrorist attacks occurred or not a week ahead in each Voronoi cell in the spatial domain (Iraq, Iran, Afghanistan, and Pakistan).

This represents 187,090 predicted cells-week over a 5-year period (2002 to 2017) using exclusively historical data. The prediction is operated incrementally using past data until the week prior to the predicted week, which ensures that historical data only is used to predict future data.

The results of the XGBoost model provides us the most important features that contribute to predict terrorism. They related to past events or characteristics of the locations of the attacks, listed here in order of importance: presence of terrorism in the previous month, occurrence of conflict, presence of terrorism, and human activity (proxied by satellite night lights) in the previous week, and the distance to the nearest country capital.

Figure 8.4 illustrates the predictive maps for week 1 (*top left*), week 10 (*top right*), week 100 (*bottom left*) and week 500 (*bottom left*) in Iraq, Iran, Pakistan, and Afghanistan.

The colors indicate the following results: (1) *green* (true positive): there is one or more attacks observed and the model correctly predicted it; (2) *white* (true negative): there is no attack and the model correctly predicted it; (3) *red* (false negative): there is one or more attacks observed but the model did not predicted the attack(s); (4) *orange* (false positive): there is no attack and the model predicted that some attack(s) occurred.

Our results show 81.8% accuracy. Accuracy is understood as the number of correct predictions divided by the total number

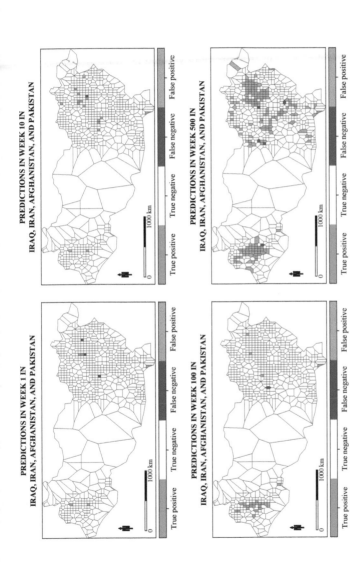

Figure 8.4: Weekly fine-scale (Voronoi polygons) predictions of terrorism illustrated for week 1 (*top left*), week 10 (*top right*), week 100 (*bottom left*) and week 500 (*bottom left*) in Iraq, Iran, Pakistan, and Afghanistan. The colors indicate the success (*green*: true positive, *white*: true negative) and failure (*red*: false negative, and *orange*: false positive) of the predictions. Terrorism data: GTD [56], covariate data [142, 103, 114, 145, 98], machine learning algorithm: XGBoost [15].

of predictions. Our model mispredicted about 21% (2,195) of the 10,315 cells that counted at least one terrorist event in a given week. In other words, about 79% of cells which encountered terrorist attacks have been accurately predicted. In addition, our model correctly predicted the absence of terrorism in 82% of the cases.

High accuracy is not a sufficient condition to ensure that the predictions are relevant for inform policy makers. The accuracy metric does not distinguish false positive from false negative. Most cells-week (176,775 cells-week over a total of 187,090 cells-week) did not encounter any terrorist attack in the investigated time period, therefore an hypothetical predictive map that would only predict the absence of terrorism in the entire study area and study period would obtain about 94% accuracy (176,775/187,090)!

In most cases, a policy-relevant approach should minimize the number of false negatives since the potential damages of terrorist events often largely exceeds the costs to prevent the attacks, especially in the case of highly lethal attacks. However, without making improvement of the models, the forced action of minimizing the false negatives leads to an increase of false positives, i.e., terrorist events that did not occur in a cell-week but are predicted as cell-week affected by terrorism.

There is therefore a trade-off between increasing the predictive ability of the ML algorithm of positive and negative events. It lies with the analyst to parametrize the ML algorithm to reflect policy-making choices, which may vary according to the available resources and employed strategy for each country and region. In our model, we added a strong penalty to make sure that the number of false negatives is strongly reduced.

8.6 Conclusion: predicting terrorism is a promising but bumpy avenue of research

We emphasized that predictions based on stochastic models differ from predictions drawn from deterministic models. Stochastic models provide probabilistic statements about a set of possible outcomes. They offer a rigorous framework to quantify uncertainty, in order to account for the randomness that is inherent to complex social phenomena such as terrorist events.

While individual acts of terrorism are virtually impossible to predict, patterns of terrorism can be identified and analyzed when one considers terrorism on an aggregated level of analysis. Progress in terrorism research remains constrained by important limitations, such as the absence of consensus on the definition of terrorism, data inaccuracy and bias, along with incomplete theory (See Section 2.2).

Despite the conceptual and methodological challenges faced by researchers in terrorism, recent development of predictive models using machine learning algorithms are promising. Our case study showed that in regions highly affected by terrorism, such as Iraq, Afghanistan or Pakistan, machine learning (ML) algorithms can provide relatively accurate predictions in both space and time and information on the most important contributors to predict terrorism that could be used to inform policy makers.

The results of this case study suggest that the use of ML algorithms may provide useful decision support tools for counterterrorism actors. We are confident that theoretical advances in the field of statistics and computer science, along with the increase of data on terrorism will allow researchers to improve the predictive performance of algorithmic models and provide practitioners active on the ground with an effective counterterrorism tool.

Galton on the central limit theorem and the Gaussian distribution

The central limit theorem states that for large sample sizes, the distribution of the sample means approximately follows a normal distribution, this, for any distribution of the data.

The power of the theorem is famously highlighted by Galton who stated:

"I know of scarcely anything so apt to impress the imagination as the wonderful form of cosmic order expressed by the law of frequency of error. The law would have been personified by the Greeks if they had known of it. It reigns with serenity and complete self-effacement amidst the wildest confusion. The larger the mob, the greater the apparent anarchy, the more perfect is its sway. It is the supreme law of unreason. Whenever a large sample of chaotic elements are taken in hand and marshaled in the order of their magnitude, an unsuspected and most beautiful form of regularity proves to have been latent all along."

Glossary

Artificial intelligence: there is no clear agreement on the definition of artificial intelligence (AI). One among many definitions of AI is the following: "a set of methods and technologies aimed at making computers or other devices function intelligently" [51].

Deterministic model: model "in which the values for the dependent variables of the system are completely determined by the parameters of the model" [120]. Deterministic models are rarely employed to model social phenomena given the presence of uncertainty in various elements (parameter, data, model specification, etc.).

Epistemology: in its narrow sense, epistemology refers to "the study of knowledge and justified belief". In its broad sense, epistemology concerns "issues having to do with the creation and dissemination of knowledge in particular areas of inquiry" [136].

Gradient boosting: is a class of machine learning algorithms that makes predictions based on average results of models added sequentially so that errors from individual models are minimized. The term "gradient" refers to descent algorithm used to minimize the loss function when new models are added [99].

Machine learning: machine learning (ML) algorithms are composed of mathematical models that use data to make predictions or decisions without being explicitly programmed to perform a specific task. One often considers ML as a branch of artificial intelligence [76].

Ontology: ontology refers to "the philosophical study of being in general, or of what applies neutrally to everything that is real" [134]. For example, some ontological questions may focus on identifying which things exist and if objective reality exists.

Predictive modeling: in statistics, predictive modeling is commonly understood as "the process of applying a statistical model or [data mining] algorithm to data for the purpose of predicting new or future observations" [133].

Stochastic models: models describing phenomena that follow a so-called stochastic process, based on laws of probability [25]. Stochastic, or probabilistic, models introduce randomness in such a way that the outcomes of the model can be viewed as probability distributions rather than unique values" [120].

While deterministic models predict a unique outcome from a given set of circumstances, stochastic models predict a set of possible outcomes through a probabilistic statement [139, p. 2].

Voronoi diagram: also called Dirichlet tessellation is composed of Voronoi polygons which are delimited by line segments. The study area is split into convex polygons so that each of them contains exactly one generating point (e.g. location of a terrorist event) and every point in a given polygon is closer to its generating point than to any other[147].

FURTHER READING

Cederman, L. E. and Weidmann, N. B. (2017). Predicting armed conflict: Time to adjust our expectations? *Science*, 355(6324), 474–476.

Cressie, N. and Wikle, C. K. (2011). Statistics for spatio-temporal data. John Wiley & Sons, Hoboken, NJ, USA, 1st Edition.

Guo, W., Gleditsch, K. and Wilson, A. (2018). Retool AI to forecast and limit wars. *Nature*, 562, 331–333.

Lichbach, M.I. (1992). Nobody cites nobody else: mathematical models of domestic political conflict. *Defence and Peace Economics*, 3(4), 341–357.

Natekin, A and Knoll, A. (2013). Gradient boosting machines, a tutorial. *Frontiers in Neurorobotics*, 7, 1–21.

Rey, S. J. (2015). Mathematical Models in Geography. In: *International Encyclopedia of the Social & Behavioral Sciences. Second Edition.*, Elsevier Inc.: 785–790.

CHAPTER 9

Terrorism: Knowns, Unknowns, and Uncertainty

Every sentence I utter must
be understood not as an
affirmation, but as a
question.

Niels Bohr

The analysis of terrorism data, mainly gathered from the Global Terrorism Database (GTD) has allowed us to debunk seven myths on terrorism. This work has brought clarity about the extent and magnitude of terrorism perpetrated by non-states across the world. In sum, it helped us distinguish what is known from what is unknown about terrorism.

Terrorism remains an ambiguous concept with disagreement among scholars and governments over the nature of the concept of terrorism. Subjectivity cannot be totally avoided and data on terrorism ineluctably reflects one among several equally reasonable interpretations of the concept.

We showed that classifying terrorism into sub-categories (e.g. civilian versus non-civilian targets) may affect the results and their interpretation. In practice, we recommend scholars to clearly state what definitions are used and acknowledge the limitations of their results accordingly.

We demystified the idea that terrorism is only about killing innocent people. From 2002 to 2017, about one half of terrorist attacks have killed people while mass-casualty events have remained very rare. In sum, few cities in the world are targeted by terrorist attacks which may or may not lead to deadly casualties.

The causes that lead terrorist groups to kill or not individuals are complex and are subject to constant scrutiny by terrorism scholars. Although few terrorist groups such as the Islamic state (ISIS) are notorious for inflicting high levels of deadly casualties, the majority of terrorist groups meticulously select their target in order to achieve their goals while minimizing their costs.

Terrorism's main targets are not in Europe, the Americas, or Oceania. From 2002 to 2017, we showed that about three quarters of the attacks worldwide took place in Asia, more particularly in Iraq, which counted more than one third of all attacks that occurred worldwide. Its capital, Baghdad, encountered circa one seventh of all lethal attacks worldwide.

Europe, the Americas, and Oceania have been far less impacted by terrorism than Asia, and to a lesser extent Africa. The realization of terrorist attacks results from complex interactions between terrorist groups and counterterrorism forces, which vary in both space and time and remain largely opaque for academic researchers and the public. However, patterns can be identified and analyzed through the use of statistical tools.

Terrorism seems to have increased globally from 2002 to 2015. The following years (2015-2017) exhibit a relative decline while the levels of terrorism remain relatively high. However, the increase observed between 2011 and 2012 is partially due to methodological inconsistencies in GTD. The reliability of statistical inference on time series data on terrorism can be improved by comparing the results based on different databases. However, there is currently no reliable alternative database that provides recent geolocated data on terrorism at fine temporal and spatial scales.

Furthermore, we notice that the observed temporal trends depend on the scale and level of aggregation in which the data are analyzed. Temporal evolution of terrorism may vary among countries, provinces and within cities. Also, one may observe differences from one year to another one, or between months, weeks, or days for example.

Terrorism is often perceived as events that occur "randomly" in space. Despite that randomness cannot be totally excluded at very fine spatial scale, the perceived randomness of terrorism is not supported by evidence above city level. Terrorist groups tend to select targets that tend to be located at close spatial proximity. Hence, terrorists often target the same or neigbhoring locations, which results in spatial clusters that can be observed at different spatial scales.

From 2002 to 2017, we observed that the majority of clusters of terrorism remained located in cities that have continuously faced high levels of terrorism, such as Baghdad or Mosul in Iraq. Using Baghdad as example, we showed that the spatial accuracy of the GTD dataset is limited, however. The current version of GTD (2018) does not allow us to capture potential spatial patterns of terrorism beyond city levels.

The spatial coverage of some clusters of terrorism change over time. Using a point process approach, we detected changes in the number and size of clusters of terrorist attacks perpetrated by ISIS in Iraq. We showed that their attacks decreased over the year in 2017 and we identified the locations affected by these changes. These dynamic mechanisms reflect changes in terrorism and/or counterterrorism activities.

In specific contexts, terrorism can be accurately predicted at relatively fine spatio-temporal scales. Here we understand the word "accurately" as sufficient enough to inform policy-makers. Predictions of terrorism remain challenging especially with regard to rare events (black swans).

We showed that stochastic models may be suitable to explain and predict terrorism since they can rigorously account for uncertainty present in terrorism data and in the modeling process. Furthermore, recent machine learning (ML) algorithms can make accurate short-term predictions in regions strongly affected by terrorism. As an illustration, we predicted terrorist attacks a week ahead at fine spatial scale in Iraq, Iran, Afghanistan and Pakistan. Our relatively high predictive performance makes ML algorithms a promising decision support tool for counterterrorism.

We wish to emphasize that one shall not attempt to generalize the results presented in this book. Despite our efforts to analyze terrorism within a large spatial and temporal spectrum, the extent to which one may generalize our results is limited. Our results are based on the Global Terrorism Database (GTD), which is currently the most comprehensive database on terrorism, but exhibits various conceptual (e.g. definition of terrorism, exclusion of state terrorism) and technical (e.g. spatial inaccuracy of events) shortcomings.

In this book, we have debunked *seven terrorism myths* through the use of statistics applied to terrorism data. This has allowed us to consolidate knowledge on terrorism while leaving various questions unanswered and raising additional doubts over the interpretation of statistical analysis of terrorism data. We hope that the development of statistical methods along with an increase of computational power and access to more comprehensive datasets will bring additional knowledge and policy-relevant guidance to better prevent and combat terrorism.

Bibliography

[1] Yonah Alexander and Dennis A Pluchinsky. *Europe's red terrorists: The fighting communist organizations*. Frank Cass and Company Ltd., 1992.

[2] Luc Anselin, Jacqueline Cohen, David Cook, Wilpen Gorr, and George Tita. Spatial analyses of crime. *Criminal justice*, 4(2):213–262, 2000.

[3] Luc Anselin and Daniel A Griffith. Do Spatial Effecfs Really Matter in Regression Analysis? *Papers in Regional Science*, 65(1):11–34, 1988.

[4] Victor H. Asal, R. Karl Rethemeyer, and Eric W. Schoon. Replication Data for: Crime, Conflict and the Legitimacy Tradeoff: Explaining Variation in Insurgents' Participation in Crime, 2018.

[5] Adrian Baddeley. Analysing spatial point patterns in R. Technical report, CSIRO, 2010. Version 4. Available at www.csiro.au/resources/pf16h.html, 2008.

[6] Adrian Baddeley. *Modeling Strategies*, pages 339–370. CRC Press, Boca Raton, FL, USA, 2010.

[7] BBC. Iraq profile - timeline, October 2018.

[8] Colin J Beck and Emily Miner. Who gets designated a terrorist and why? *Social Forces*, 91:837–872, 2013.

[9] Brandon Behlendorf, Gary LaFree, and Richard Legault. Microcycles of violence: Evidence from terrorist attacks by ETA and the FMLN. *Journal of Quantitative Criminology*, 28:49–75, 2012.

[10] Nathan Black. When have violent civil conflicts spread? Introducing a dataset of substate conflict contagion. *Journal of Peace Research*, 50(6):751–759, 2013.

[11] Alex Braithwaite and Quan Li. Transnational terrorism hot spots: Identification and impact evaluation. *Conflict Management and Peace Science*, 24:281–296, 2007.

[12] CEACS. Explaining terrorist and insurgent behavior. `http://www.march.es/ceacs/proyectos/dtv/datasets.asp`, 2013. [Accessed 12 June 2014].

[13] Lars-Erik Cederman and Nils B Weidmann. Predicting armed conflict: Time to adjust our expectations? *Science*, 355(6324):474–476, 2017.

[14] Center for International Earth Science Information Network (CIESIN) Columbia University. Global Rural-Urban Mapping Project, Version 1 (GRUMPv1): Urban Extent Polygons. *NASA Socioeconomic Data and Applications Center (SEDAC)*, 2018.

[15] Tianqi Chen and Carlos Guestrin. Xgboost: A scalable tree boosting system. In *Proceedings of the 22nd ACM SIGKDD international conference on knowledge discovery and data mining*, pages 785–794. ACM, 2016.

[16] Noam Chomsky. *Pirates and Emperors, Old and New: International Terrorism in the Real World*. Pluto Press, new edition edition, 2002.

[17] Noam Chomsky. *Power and Terror: Post-9/11 Talks and Interviews*. Seven Stories Press, New York, junkerman, john and masakazu, takei edition, 2003.

[18] Noam Chomsky. *Failed States: The Abuse of Power and The Assault on Democracy*. Penguin Books, 2006.

[19] CIA. The world factbook: Pakistan. Technical report, Central Intelligence Agency (CIA), Office of Public Affairs, Washington, D.C. Available at `https://www.cia.gov/library/publications/the-world-factbook/fields/397.html#PK`, 2017.

[20] Aaron Clauset, Maxwell Young, and Kristian S Gleditsch. On the frequency of severe terrorist events. *Journal of Conflict Resolution*, 51:58–87, 2007.

[21] Christina Cliff and Andrew First. Testing for contagion/diffusion of terrorism in state dyads. *Studies in Conflict & Terrorism*, 36(4):292–314, 2013.

[22] COBUILD Advanced English Dictionary. "modus operandi, n." HarperCollins Publishers Limited, November 2019. [Accessed 19 November 2019].

[23] Jacqueline Cohen and George Tita. Diffusion in homicide: Exploring a general method for detecting spatial diffusion processes. *Journal of Quantitative Criminology*, 15(4):451–493, 1999.

[24] Copeland, B.J. for Encyclopaedia Britannica Online. "artificial intelligence". https://www.britannica.com/technology/artificial-intelligence, May 2019. [Accessed 25 september 2019].

[25] David Roxbee Cox. *The theory of stochastic processes*. Routledge, 2017.

[26] Ronald Crelinsten. *Counterterrorism*. Polity Press, Cambridge, UK, 2009.

[27] Martha Crenshaw. The causes of terrorism. *Comparative Politics*, 13:379–399, 1981.

[28] Martha Crenshaw. *Introduction: Reflections on the Effects of Terrorism*, pages 1–37. Wesleyan University Press, Middletown, CT, USA, 1983.

[29] Martha Crenshaw. *The Causes of Terrorism*, pages 113–126. St. Martin's Press, Inc, New York, NY, USA, 1990.

[30] Martha Crenshaw. How terrorism declines. *Terrorism and Political Violence*, 3:69–87, 1991.

[31] Noel Cressie and Christopher K Wikle. *Statistics for Spatio-Temporal Data*. John Wiley & Sons, Hoboken, NJ, USA, 1st edition, 2011.

[32] Audrey Kurth Cronin. Behind the curve: Globalization and international terrorism. *International Security*, 27:30–58, 2003.

[33] Peter J Diggle. *Statistical Analysis of Spatial and Spatio-Temporal Point Patterns*. CRC Press, Boca Raton, FL, USA, 3rd edition, 2014.

[34] Fangyu Ding, Quansheng Ge, Dong Jiang, Jingying Fu, and Mengmeng Hao. Understanding the dynamics of terrorism events with multiple-discipline datasets and machine learning approach. *Plos one*, 12(6):e0179057, 2017.

[35] Karsten Donnay, Eric T Dunford, Erin C McGrath, David Backer, and David E Cunningham. Integrating conflict event data. *Journal of Conflict Resolution*, 63(5):1337–1364, 2019.

[36] Charles John Michael Drake. *Terrorists' Target Selection*. St. Martin's Press, New York, NY, USA, 1st edition, 1998.

[37] Konstantinos Drakos. The size of under-reporting bias in recorded transnational terrorist activity. *Journal of the Royal Statistical Society: Series A (Statistics in Society)*, 170(4):909–921, 2007.

[38] Konstantinos Drakos and Andreas Gofas. In search of the average transnational terrorist attack venue. *Defence and Peace Economics*, 17:73–93, 2006.

[39] Christopher D Elvidge, Jeffrey Safran, Benjamin Tuttle, Paul Sutton, Pierantonio Cinzano, Donald Pettit, John Arvesen, and Christopher Small. Potential for global mapping of development via a nightsat mission. *GeoJournal*, 69(1-2):45–53, 2007.

[40] Encyclopaedia Britannica Online. "diffusion". Encyclopaedia Britannica Inc., December 2014. [Accessed 17 December 2014].

[41] Walter Enders and Todd Sandler. Transnational terrorism in the post-cold war era. *International Studies Quarterly*, 43:145–167, 1999.

[42] Walter Enders and Todd Sandler. Is international terrorism becoming more threatening? A time series investigation. *Journal of Conflict Resolution*, 44:307–322, 2000.

[43] Walter Enders and Todd Sandler. Transnational terrorism 1968-2000: Thresholds, persistence and forecasts. *Southern Economic Journal*, 71:467–482, 2005.

[44] Walter Enders and Todd Sandler. Distribution of transnational terrorism among countries by income class and geography after 9/11. *International Studies Quarterly*, 50(2):367–393, 2006.

[45] Walter Enders, Todd Sandler, and Khusrav Gaibulloev. Domestic versus transnational terrorism: Data, decomposition, and dynamics. *Journal of Peace Research*, 48:319–337, 2011.

[46] Jan Oskar Engene. Five decades of terrorism in Europe: The TWEED dataset. *Journal of Peace Research*, 44:109–121, 2007.

[47] Richard English. *Terrorism: How to respond.* Oxford University Press, 2010.

[48] ESRI. Arcgis desktop: Release 10.6.1. environmental systems research institute. redlands, ca. Software, 2019. ESRI 2019. ArcGIS Desktop: Release 10.

[49] James D. Fearon. Civil war & the current international system. *Daedalus*, 146(4):18–32, 2017.

[50] Erika Forsberg. Diffusion in the study of civil wars: A cautionary tale. *International Studies Review*, 16:188–198, 2014.

[51] G7 science academies. Artificial intelligence and society. Summit of the G7 Science Academies. March 25-26, 2019. https://royalsociety.org/-/media/about-us/international/g-science-statements/2019-g7-declaration-artificial-intelligence-and-society.pdf?la=en-GB&hash=0F5AC7386F43088FF9A5E55BBB3E56BE, 2019. [Accessed 26 September 2019].

[52] Peng Gao, Diansheng Guo, Ke Liao, Jennifer J Webb, and Susan L Cutter. Early Detection of Terrorism Outbreaks Using Prospective Space-Time Scan Statistics. *The Professional Geographer*, 65:676–691, 2013.

[53] Jack P Gibbs. Conceptualization of terrorism. *American Sociological Review*, pages 329–340, 1989.

[54] John M. Gleason. A Poisson model of incidents of international terrorism in the United States. *Terrorism*, 4:259–265, 1980.

[55] Jean-Germain Gros. Towards a taxonomy of failed states in the New World Order: decaying Somalia, Liberia, Rwanda and Haiti. *Third World Quarterly*, 17(3):455–472, 1996.

[56] GTD. Global Terrorism Database (GTD) Codebook: Inclusion Criteria and Variables. http://www.start.umd.edu/gtd/downloads/Codebook.pdf, August 2018. [Accessed 9 December 2018].

[57] Weisi Guo, Kristian Gleditsch, and Alan Wilson. Retool AI to forecast and limit wars. *Nature (Comment)*, 562:331–333, 2018.

[58] James Douglas Hamilton. *Time Series Analysis*. Princeton University Press, Princeton, NJ, USA, 1st edition, 1994.

[59] Lawrence C Hamilton and James D Hamilton. Dynamics of terrorism. *International Studies Quarterly*, 27:39–54, 1983.

[60] Andrew C Harvey. *Time Series Models*. MIT Press, Cambridge, MA, USA, 2nd edition, 1993.

[61] Gerald B Helman and Steven R Ratner. Saving failed states. *Foreign policy*, 89(3):3–20, 1992.

[62] Edward Heyman and Edward Mickolus. Observations on "why violence spreads". *International Studies Quarterly*, 24:299–305, 1980.

[63] Bruce Hoffman. *Inside Terrorism*. Columbia University Press, New York, NY, USA, rev. and expanded edition, 2006.

[64] Rob J Hyndman. Moving averages. In *International Encyclopedia of Statistical Science*, pages 866–869. Springer, 2011.

[65] Janine Illian, Antti Penttinen, Helga Stoyan, and Dietrich Stoyan. *Statistical Analysis and Modelling of Spatial Point patterns*. John Wiley & Sons Ltd, Chichester, West Sussex, UK, 1st edition, 2008.

[66] International Committee of the Red Cross (ICRC). Geneva Convention Relative to the Protection of Civilian Persons in Time of War (Fourth Geneva Convention). 75 UNTS 287, August 1949.

[67] International Committee of the Red Cross (ICRC). Protocol Additional to the Geneva Conventions of 12 August 1949. 1125 UNTS 3, June 1977.

[68] International Court of Justice. Military and Paramilitary Activities in and against Nicaragua (Nicaragua v. United States of America). Judgment of 27 June 1986. http://www.icj-cij.org/docket/?sum=367&p1=3&p2=3&case=70&p3=5, 1986. [Accessed 26 August 2015].

[69] Valerie Isham. *Spatial Point Process Models*, pages 283–298. CRC Press, Boca Raton, FL, USA, 2010.

[70] Brian M Jenkins and Janera Johnson. International terrorism: A chronology, 1968-1974. Technical report, DTIC Document, Santa Monica, CA, USA, 1975.

[71] John Junkerman. Power and terror: Noam Chomsky in our times. Film directed by John Junkerman, 2002. Documentary.

[72] Daniel Kahneman. *Thinking, Fast And Slow*. Penguin Books, London, UK, 1st edition, 2011.

[73] Stephen H Kellert. *In the wake of chaos: Unpredictable order in dynamical systems*. University of Chicago press, 1994.

[74] John Frank Charles Kingman. *Poisson processes*. Oxford university press, Oxford, United Kingdom, 1st edition, 1992.

[75] Judy L Klein. *Statistical Visions in Time: A History of Time Series Analysis, 1662-1938*. Cambridge University Press, Cambridge, UK, 1st edition, 1997.

[76] John R Koza, Forrest H Bennett, David Andre, and Martin A Keane. Automated design of both the topology and sizing of analog electrical circuits using genetic programming. In *Artificial Intelligence in Design'96*, pages 151–170. Springer, 1996.

[77] Alan B Krueger. *What Makes a Terrorist: Economics and the Roots of Terrorism*. Princeton Unversity Press, Princeton, NJ, USA, 2017.

[78] Minjung Kyung, Jeff Gill, and George Casella. New findings from terrorism data: Dirichlet process random-effects models for latent groups. *Journal of the Royal Statistical Society. Series C. Applied Statistics*, 60:701–721, 2011.

[79] Gary LaFree and Laura Dugan. Introducing the Global Terrorism Database. *Terrorism and Political Violence*, 19(2):181–204, 2007.

[80] Gary LaFree, Laura Dugan, and Erin Miller. *Putting terrorism in context: Lessons from the Global Terrorism Database.* Routledge, 2014.

[81] Gary LaFree, Laura Dugan, Min Xie, and Piyusha Singh. Spatial and temporal patterns of terrorist attacks by ETA 1970 to 2007. *Journal of Quantitative Criminology*, 28:7–29, 2012.

[82] Gary LaFree, Nancy A Morris, and Laura Dugan. Cross-national patterns of terrorism comparing trajectories for total, attributed and fatal attacks, 1970–2006. *British Journal of Criminology*, 50(4):622–649, 2010.

[83] Gary LaFree, Sue-Ming Yang, and Martha Crenshaw. Trajectories of terrorism. *Criminology & Public Policy*, 8(3):445–473, 2009.

[84] Kalev Leetaru and Philip A Schrodt. GDELT: Global data on events, location, and tone, 1979–2012. In *Paper presented at the ISA Annual Convention*, volume 2, page 4, 2013.

[85] Erik Lewis, George Moehler, P Jeffrey Brantingham, and Andrea L Bertozzi. Self-exciting point process models of civilian deaths in Iraq. *Security Journal*, 25:244–264, 2012.

[86] Mark I Lichbach. Nobody cites nobody else: Mathematical models of domestic political conflict. *Defence and Peace Economics*, 3(4):341–357, 1992.

[87] May Lim, Richard Metzler, and Yaneer Bar-Yam. Global pattern formation and ethnic/cultural violence. *Science*, 317(5844):1540–1544, 2007.

[88] Andrew M Linke, Frank DW Witmer, and John O'Loughlin. Space-time granger analysis of the war in Iraq: A study of coalition and insurgent action-reaction. *International Interactions*, 38(4):402–425, 2012.

[89] Allan Mazur. Bomb threats and the mass media: Evidence for a theory of suggestion. *American Sociological Review*, 47:407–411, 1982.

[90] Kathleen McKendrick. Artificial intelligence prediction and counterterrorism. Technical report, Research Paper. International Security Department, Chatham House, The Royal Institute of International Affairs. August 2019. Available at https://www.chathamhouse.org/sites/default/files/2019-08-07-AICounterterrorism.pdf, 2019.

[91] Richard M Medina, Laura K. Siebeneck, and George F Hepner. A geographic information systems (GIS) analysis of spatiotemporal patterns of terrorist incidents in Iraq 2004-2009. *Studies in Conflict & Terrorism*, 34:862–882, 2011.

[92] Edward Mickolus. *International terrorism attributes of terrorist events (ITERATE)*. Vinyard Software, Dunn Loring, VA, USA, 2003. Data codebook.

[93] Manus I Midlarsky, Martha Crenshaw, and Fumihiko Yoshida. Why violence spreads: The contagion of international terrorism. *International Studies Quarterly*, 24:262–298, 1980.

[94] George Mohler. Modeling and estimation of multi-source clustering in crime and security data. *The Annals of Applied Statistics*, 7:1525–1539, 2013.

[95] José G Montalvo and Marta Reynal-Querol. Ethnic polarization, potential conflict, and civil wars. *American economic review*, 95(3):796–816, 2005.

[96] John Mueller. Six rather unusual propositions about terrorism. *Terrorism and Political Violence*, 17:487–505, 2005.

[97] Alexander B Murphy. The space of terror. In S. l. Cutter, D. B. Richardson, and T. J. Wilbanks, editors, *The Geographical Dimension of Terrorism*, chapter The Space of Terror, pages 47–52. Routledge, 2003.

[98] NASA Socioeconomic Data and Applications Center (SEDAC), Palisades NY, USA. Gridded Population of the World, Version 4.10 data sets (GPWv4). https://doi.org/10.7927/H4B56GPT, 2017. [Accessed 12 August 2018].

[99] Alexey Natekin and Alois Knoll. Gradient boosting machines, a tutorial. *Frontiers in Neurorobotics*, 7:21, 2013.

[100] Michael Nauenberg. *Isaac Newton's Natural Philosophy.* MIT Press, 2001.

[101] Stephen C. Nemeth, Jacob A. Mauslein, and Craig Stapley. The primacy of the local: Identifying terrorist hot spots using Geographic Information Systems. *The Journal of Politics,* 76:304–317, 4 2014.

[102] Eric Neumayer and Thomas Plümper. Galton's problem and contagion in international terrorism along civilizational lines. *Conflict management and peace science,* 27(4):308–325, 2010.

[103] NOAA. *Version 4 DMSP-OLS Nighttime Lights Time Series.* National Oceanic and Atmospheric Administration, National Geophysical Data Center, 2014.

[104] Samuel Nunn. Incidents of terrorism in the United States, 1997-2005. *Geographical Review,* 97:89–111, 2007.

[105] Conor Cruise O'Brien. *Terrrorism under Democratic Conditions: The Case of the IRA,* pages 91–104. Wesleyan University Press, Middletown, CT, USA, 1983.

[106] OpenStreetMap. ArcGIS Online background map. World light gray canvas base map. Software, 2019.

[107] Oxford English Dictionary Online. "diffusion, n." Oxford University Press, December 2014. [Accessed 17 December 2014].

[108] Robert A Pape. The strategic logic of suicide terrorism. *American Political Science Review,* 97(03):343–361, 2003.

[109] Walter W. Piegorsch, Susan L. Cutter, and Frank Hardisty. Benchmark analysis for quantifying urban vulnerability to terrorrst incidents. *Risk Analysis,* 27:1411–1425, 2007.

[110] David Pion-Berlin and George A. Lopez. Of victims and executioners: Argentine state terror, 1975-1979. *International Studies Quarterly,* 35(1):63–86, 1991.

[111] Michael D. Porter and Gentry White. Self-exciting hurdle models for terrorist activity. *The Annals of Applied Statistics,* 6(1):106–124, 2012.

[112] PSU. Lesson 5.1: Decomposition models.STAT 510: Applied time series analysis. https://onlinecourses.science. psu.edu/stat510/node/69, October 2014.

[113] Andre Python. *Modelling the Spatial Dynamics of Terrorism: World Study, 2002-2013*. PhD thesis, School of Mathematics and Statistics, University of St Andrews, United Kingdom, 2017. Unpublished thesis.

[114] Andre Python, Jürgen Brandsch, and Aliya Tskhay. Provoking local ethnic violence–a global study on ethnic polarization and terrorist targeting. *Political Geography*, 58:77–89, 2017.

[115] André Python, Janine Illian, Charlotte Jones-Todd, and Marta Blangiardo. A Bayesian Approach to Modelling Subnational Spatial Dynamics of Worldwide Non-State Terrorism, 2010 - 2016. *Journal of the Royal Statistical Society - Series A : Statistics in Society*, 182(1):323–344, 2019.

[116] R Core Team. R: A language and environment for statistical computing. http://www.R-project.org/, 2014. [Accessed 2 December 2014].

[117] Vasanthan Raghavan, Aram Galstyan, and Alexander G Tartakovsky. Hidden Markov models for the activity profile of terrorist groups. *The Annals of Applied Statistics*, 7(4):2402–2430, 2013.

[118] RAND. Database scope. http://www.rand.org/nsrd/projects/terrorism-incidents/about/scope.html, 2011. [Accessed 12 June 2014].

[119] David C. Rapoport. Editorial: The media and terrorism: Implications of the Unabomber case. *Terrorism and Political Violence*, 8(1):7–9, 1996.

[120] Sergio J Rey. Mathematical models in geography. In *International Encyclopedia of the Social & Behavioral Sciences: Second Edition*, pages 785–790. Elsevier Inc., 2015.

[121] Marta Reynal-Querol and Jose G Montalvo. Ethnic polarization, potential conflict and civil war. *American Economic Review*, 95(3):796–816, 2005.

[122] Louise Richardson. *What Terrorists Want*. John Murray, London, UK, 1st edition, 2006.

[123] Alexandre Rodrigues, Peter Diggle, and Renato Assuncao. Semiparametric approach to point source modelling in epidemiology and criminology. *Journal of the Royal Statistical Society - Series C Applied Statistics*, 59:533–542, 2010.

[124] Idean Salehyan. *Rebels Without Borders: States Boundaries, Transnational Opposition, and Civil Conflict.* PhD thesis, University of San Diego, USA, 2006.

[125] Nicholas Sambanis. Terrorism and civil war. In Philip Keefer and Norman Loayza, editors, *Terrorism, Economic Development, and Political Openness*, pages 174–206. Cambridge University Press, New York, NY, USA, 2008.

[126] Todd Sandler and Walter Enders. Applying analytical methods to study terrorism. *International Studies Perspectives*, 8(3):287–302, 2007.

[127] Alan P Schmid and Albert J Jongman. *Political Terrorism: A New Guide to Actors, Authors, Concepts, Data Bases, Theories, and Literature.* Transaction Books, New Brunswick, NJ, USA, 2nd edition, 1988.

[128] Alex P Schmid. Terrorism and democracy. *Terrorism and Political Violence*, 4(4):14–25, 1992.

[129] Alex P Schmid. The revised academic consensus definition of terrorism. *Perspectives on Terrorism*, 6(2), 2012.

[130] Erwin Schrödinger. Quantisierung als Eigenwertproblem. *Annalen der Physik*, 385(13):437–490, 1926.

[131] Sebastian Schutte and Nils B Weidmann. Diffusion patterns of violence in civil wars. *Political Geography*, 30(3):143–152, 2011.

[132] Ivan Sascha Sheehan. *Assessing and Comparing Data Sources for Terrorism Research*, pages 13–40. Springer, New York, NY, USA, 2012.

[133] Galit Shmueli et al. To explain or to predict? *Statistical science*, 25(3):289–310, 2010.

[134] P.M. Simons. Ontology. https://www.britannica.com/topic/ontology-metaphysics, 2015.

[135] Jessica A. Stanton. Terrorism in the context of civil war. *The Journal of Politics*, 75(4):1009–1022, 2013.

[136] Matthias Steup. Epistemology. https://plato.stanford. edu/archives/win2018/entries/epistemology, 2018.

[137] Ralph Sundberg and Erik Melander. Introducing the ucdp georeferenced event dataset. *Journal of Peace Research*, 50(4):523–532, 2013.

[138] Paul C Sutton, Christopher D Elvidge, and Tilottama Ghosh. Estimation of gross domestic product at sub-national scales using nighttime satellite imagery. *International Journal of Ecological Economics & Statistics*, 8(S07):5–21, 2007.

[139] Howard M Taylor, editor. *An Introduction to Stochastic Modeling*. Academic Press, 3rd edition edition, 2014.

[140] The New York Times. Boko Haram is back. With better drones, September 2019.

[141] Waldo R Tobler. A computer movie simulating urban growth in the Detroit region. *Economic geography*, pages 234–240, 1970.

[142] Andreas Forø Tollefsen, Håvard Strand, and Halvard Buhaug. PRIO-GRID: A unified spatial data structure. *Journal of Peace Research*, 49(2):363–374, 2012.

[143] Peter Turchin, Thomas E Currie, Edward AL Turner, and Sergey Gavrilets. War, space, and the evolution of old world complex societies. *Proceedings of the National Academy of Sciences*, 110(41):16384–16389, 2013.

[144] Leonard Weinberg, Ami Pedahzur, and Sivan Hirsch-Hoefler. The challenges of conceptualizing terrorism. *Terrorism and Policical Violence*, 16(4):777–794, 2004.

[145] DJ Weiss, A Nelson, HS Gibson, W Temperley, S Peedell, A Lieber, M Hancher, E Poyart, S Belchior, N Fullman, et al. A global map of travel time to cities to assess inequalities in accessibility in 2015. *Nature*, 553(7688):333, 2018.

[146] Eric W. Weisstein. Buffon's Needle Problem. From MathWorld–A Wolfram Web Resource. http://mathworld.wolfram.com/BuffonsNeedleProblem.html, 2019. [Accessed 1 November 2019].

[147] Eric W. Weisstein. Voronoi Diagram. From MathWorld–A Wolfram Web Resource. http://mathworld.wolfram.com/ VoronoiDiagram.html, 2019. [Accessed 1 December 2019].

[148] Paul Wilkinson. *Terrorism and the Liberal State.* The MacMillan Press Ltd, Hong Kong, China, 2nd edition, 1979.

[149] Wilson Center. Timeline: the rise, spread, and fall of the Islamic State, April 2019.

[150] Andrew Zammit Mangion. *Modelling from spatiotemporal data: a dynamic systems approach.* PhD thesis, University of Sheffield, UK, 2011.

[151] Andrew Zammit-Mangion, Michael Dewar, Visan Kadirka-manathan, Anaid Flesken, and Guido Sanguinetti. *Modeling Conflict Dynamics with Spatio-temporal Data.* SpringerBriefs in Applied Sciences and Technology, Heidelberg, Germany, 1st edition, 2013.

[152] Andrew Zammit-Mangion, Michael Dewar, Visakan Kadirka-manathand, and Guido Sanguinetti. Point Process Modelling of the Afghan war diary. *PNAS*, 109:12414–12419, 2012.

[153] I William Zartman. Introduction: Posing the Problem of State Collapse. In *Collapsed states: the disintegration and restoration of legitimate authority*, pages 1–13. Lynne Rienner Publishers, 1995.

Index

Milton Keynes UK
Ingram Content Group UK Ltd.
UKHW051011071024
449327UK00001B/2